Implementing Concurrent Engineering in Small Companies

MECHANICAL ENGINEERING
A Series of Textbooks and Reference Books

Founding Editor

L. L. Faulkner

Columbus Division, Battelle Memorial Institute
and Department of Mechanical Engineering
The Ohio State University
Columbus, Ohio

1. *Spring Designer's Handbook*, Harold Carlson
2. *Computer-Aided Graphics and Design*, Daniel L. Ryan
3. *Lubrication Fundamentals*, J. George Wills
4. *Solar Engineering for Domestic Buildings*, William A. Himmelman
5. *Applied Engineering Mechanics: Statics and Dynamics*, G. Boothroyd and C. Poli
6. *Centrifugal Pump Clinic*, Igor J. Karassik
7. *Computer-Aided Kinetics for Machine Design*, Daniel L. Ryan
8. *Plastics Products Design Handbook, Part A: Materials and Components; Part B: Processes and Design for Processes*, edited by Edward Miller
9. *Turbomachinery: Basic Theory and Applications*, Earl Logan, Jr.
10. *Vibrations of Shells and Plates*, Werner Soedel
11. *Flat and Corrugated Diaphragm Design Handbook*, Mario Di Giovanni
12. *Practical Stress Analysis in Engineering Design*, Alexander Blake
13. *An Introduction to the Design and Behavior of Bolted Joints*, John H. Bickford
14. *Optimal Engineering Design: Principles and Applications*, James N. Siddall
15. *Spring Manufacturing Handbook*, Harold Carlson
16. *Industrial Noise Control: Fundamentals and Applications*, edited by Lewis H. Bell
17. *Gears and Their Vibration: A Basic Approach to Understanding Gear Noise*, J. Derek Smith
18. *Chains for Power Transmission and Material Handling: Design and Applications Handbook*, American Chain Association
19. *Corrosion and Corrosion Protection Handbook*, edited by Philip A. Schweitzer
20. *Gear Drive Systems: Design and Application*, Peter Lynwander
21. *Controlling In-Plant Airborne Contaminants: Systems Design and Calculations*, John D. Constance
22. *CAD/CAM Systems Planning and Implementation*, Charles S. Knox
23. *Probabilistic Engineering Design: Principles and Applications*, James N. Siddall
24. *Traction Drives: Selection and Application*, Frederick W. Heilich III and Eugene E. Shube
25. *Finite Element Methods: An Introduction*, Ronald L. Huston and Chris E. Passerello

26. *Mechanical Fastening of Plastics: An Engineering Handbook*, Brayton Lincoln, Kenneth J. Gomes, and James F. Braden
27. *Lubrication in Practice: Second Edition*, edited by W. S. Robertson
28. *Principles of Automated Drafting*, Daniel L. Ryan
29. *Practical Seal Design*, edited by Leonard J. Martini
30. *Engineering Documentation for CAD/CAM Applications*, Charles S. Knox
31. *Design Dimensioning with Computer Graphics Applications*, Jerome C. Lange
32. *Mechanism Analysis: Simplified Graphical and Analytical Techniques*, Lyndon O. Barton
33. *CAD/CAM Systems: Justification, Implementation, Productivity Measurement*, Edward J. Preston, George W. Crawford, and Mark E. Coticchia
34. *Steam Plant Calculations Manual*, V. Ganapathy
35. *Design Assurance for Engineers and Managers*, John A. Burgess
36. *Heat Transfer Fluids and Systems for Process and Energy Applications*, Jasbir Singh
37. *Potential Flows: Computer Graphic Solutions*, Robert H. Kirchhoff
38. *Computer-Aided Graphics and Design: Second Edition*, Daniel L. Ryan
39. *Electronically Controlled Proportional Valves: Selection and Application*, Michael J. Tonyan, edited by Tobi Goldoftas
40. *Pressure Gauge Handbook*, AMETEK, U.S. Gauge Division, edited by Philip W. Harland
41. *Fabric Filtration for Combustion Sources: Fundamentals and Basic Technology*, R. P. Donovan
42. *Design of Mechanical Joints*, Alexander Blake
43. *CAD/CAM Dictionary*, Edward J. Preston, George W. Crawford, and Mark E. Coticchia
44. *Machinery Adhesives for Locking, Retaining, and Sealing*, Girard S. Haviland
45. *Couplings and Joints: Design, Selection, and Application*, Jon R. Mancuso
46. *Shaft Alignment Handbook*, John Piotrowski
47. *BASIC Programs for Steam Plant Engineers: Boilers, Combustion, Fluid Flow, and Heat Transfer*, V. Ganapathy
48. *Solving Mechanical Design Problems with Computer Graphics*, Jerome C. Lange
49. *Plastics Gearing: Selection and Application*, Clifford E. Adams
50. *Clutches and Brakes: Design and Selection*, William C. Orthwein
51. *Transducers in Mechanical and Electronic Design*, Harry L. Trietley
52. *Metallurgical Applications of Shock-Wave and High-Strain-Rate Phenomena*, edited by Lawrence E. Murr, Karl P. Staudhammer, and Marc A. Meyers
53. *Magnesium Products Design*, Robert S. Busk
54. *How to Integrate CAD/CAM Systems: Management and Technology*, William D. Engelke
55. *Cam Design and Manufacture: Second Edition*; with cam design software for the IBM PC and compatibles, disk included, Preben W. Jensen
56. *Solid-State AC Motor Controls: Selection and Application*, Sylvester Campbell
57. *Fundamentals of Robotics*, David D. Ardayfio
58. *Belt Selection and Application for Engineers*, edited by Wallace D. Erickson
59. *Developing Three-Dimensional CAD Software with the IBM PC*, C. Stan Wei
60. *Organizing Data for CIM Applications*, Charles S. Knox, with contributions by Thomas C. Boos, Ross S. Culverhouse, and Paul F. Muchnicki

61. *Computer-Aided Simulation in Railway Dynamics*, by Rao V. Dukkipati and Joseph R. Amyot
62. *Fiber-Reinforced Composites: Materials, Manufacturing, and Design*, P. K. Mallick
63. *Photoelectric Sensors and Controls: Selection and Application*, Scott M. Juds
64. *Finite Element Analysis with Personal Computers*, Edward R. Champion, Jr., and J. Michael Ensminger
65. *Ultrasonics: Fundamentals, Technology, Applications: Second Edition, Revised and Expanded*, Dale Ensminger
66. *Applied Finite Element Modeling: Practical Problem Solving for Engineers*, Jeffrey M. Steele
67. *Measurement and Instrumentation in Engineering: Principles and Basic Laboratory Experiments*, Francis S. Tse and Ivan E. Morse
68. *Centrifugal Pump Clinic: Second Edition, Revised and Expanded*, Igor J. Karassik
69. *Practical Stress Analysis in Engineering Design: Second Edition, Revised and Expanded*, Alexander Blake
70. *An Introduction to the Design and Behavior of Bolted Joints: Second Edition, Revised and Expanded*, John H. Bickford
71. *High Vacuum Technology: A Practical Guide*, Marsbed H. Hablanian
72. *Pressure Sensors: Selection and Application*, Duane Tandeske
73. *Zinc Handbook: Properties, Processing, and Use in Design*, Frank Porter
74. *Thermal Fatigue of Metals*, Andrzej Weronski and Tadeusz Hejwowski
75. *Classical and Modern Mechanisms for Engineers and Inventors*, Preben W. Jensen
76. *Handbook of Electronic Package Design*, edited by Michael Pecht
77. *Shock-Wave and High-Strain-Rate Phenomena in Materials*, edited by Marc A. Meyers, Lawrence E. Murr, and Karl P. Staudhammer
78. *Industrial Refrigeration: Principles, Design and Applications*, P. C. Koelet
79. *Applied Combustion*, Eugene L. Keating
80. *Engine Oils and Automotive Lubrication*, edited by Wilfried J. Bartz
81. *Mechanism Analysis: Simplified and Graphical Techniques, Second Edition, Revised and Expanded*, Lyndon O. Barton
82. *Fundamental Fluid Mechanics for the Practicing Engineer*, James W. Murdock
83. *Fiber-Reinforced Composites: Materials, Manufacturing, and Design, Second Edition, Revised and Expanded*, P. K. Mallick
84. *Numerical Methods for Engineering Applications*, Edward R. Champion, Jr.
85. *Turbomachinery: Basic Theory and Applications, Second Edition, Revised and Expanded*, Earl Logan, Jr.
86. *Vibrations of Shells and Plates: Second Edition, Revised and Expanded*, Werner Soedel
87. *Steam Plant Calculations Manual: Second Edition, Revised and Ex panded*, V. Ganapathy
88. *Industrial Noise Control: Fundamentals and Applications, Second Edition, Revised and Expanded*, Lewis H. Bell and Douglas H. Bell
89. *Finite Elements: Their Design and Performance*, Richard H. MacNeal
90. *Mechanical Properties of Polymers and Composites: Second Edition, Revised and Expanded*, Lawrence E. Nielsen and Robert F. Landel
91. *Mechanical Wear Prediction and Prevention*, Raymond G. Bayer

92. *Mechanical Power Transmission Components*, edited by David W. South and Jon R. Mancuso
93. *Handbook of Turbomachinery*, edited by Earl Logan, Jr.
94. *Engineering Documentation Control Practices and Procedures*, Ray E. Monahan
95. *Refractory Linings Thermomechanical Design and Applications*, Charles A. Schacht
96. *Geometric Dimensioning and Tolerancing: Applications and Techniques for Use in Design, Manufacturing, and Inspection*, James D. Meadows
97. *An Introduction to the Design and Behavior of Bolted Joints: Third Edition, Revised and Expanded*, John H. Bickford
98. *Shaft Alignment Handbook: Second Edition, Revised and Expanded*, John Piotrowski
99. *Computer-Aided Design of Polymer-Matrix Composite Structures*, edited by Suong Van Hoa
100. *Friction Science and Technology*, Peter J. Blau
101. *Introduction to Plastics and Composites: Mechanical Properties and Engineering Applications*, Edward Miller
102. *Practical Fracture Mechanics in Design*, Alexander Blake
103. *Pump Characteristics and Applications*, Michael W. Volk
104. *Optical Principles and Technology for Engineers*, James E. Stewart
105. *Optimizing the Shape of Mechanical Elements and Structures*, A. A. Seireg and Jorge Rodriguez
106. *Kinematics and Dynamics of Machinery*, Vladimír Stejskal and Michael Valášek
107. *Shaft Seals for Dynamic Applications*, Les Horve
108. *Reliability-Based Mechanical Design*, edited by Thomas A. Cruse
109. *Mechanical Fastening, Joining, and Assembly*, James A. Speck
110. *Turbomachinery Fluid Dynamics and Heat Transfer*, edited by Chunill Hah
111. *High-Vacuum Technology: A Practical Guide, Second Edition, Revised and Expanded*, Marsbed H. Hablanian
112. *Geometric Dimensioning and Tolerancing: Workbook and Answerbook*, James D. Meadows
113. *Handbook of Materials Selection for Engineering Applications*, edited by G. T. Murray
114. *Handbook of Thermoplastic Piping System Design*, Thomas Sixsmith and Reinhard Hanselka
115. *Practical Guide to Finite Elements: A Solid Mechanics Approach*, Steven M. Lepi
116. *Applied Computational Fluid Dynamics*, edited by Vijay K. Garg
117. *Fluid Sealing Technology*, Heinz K. Muller and Bernard S. Nau
118. *Friction and Lubrication in Mechanical Design*, A. A. Seireg
119. *Influence Functions and Matrices*, Yuri A. Melnikov
120. *Mechanical Analysis of Electronic Packaging Systems*, Stephen A. McKeown
121. *Couplings and Joints: Design, Selection, and Application, Second Edition, Revised and Expanded,* Jon R. Mancuso
122. *Thermodynamics: Processes and Applications*, Earl Logan, Jr.
123. *Gear Noise and Vibration*, J. Derek Smith
124. *Practical Fluid Mechanics for Engineering Applications,* John J. Bloomer
125. *Handbook of Hydraulic Fluid Technology*, edited by George E. Totten
126. *Heat Exchanger Design Handbook*, T. Kuppan

127. *Designing for Product Sound Quality*, Richard H. Lyon
128. *Probability Applications in Mechanical Design*, Franklin E. Fisher and Joy R. Fisher
129. *Nickel Alloys,* edited by Ulrich Heubner
130. *Rotating Machinery Vibration: Problem Analysis and Troubleshooting*, Maurice L. Adams, Jr.
131. *Formulas for Dynamic Analysis,* Ronald L. Huston and C. Q. Liu
132. *Handbook of Machinery Dynamics*, Lynn L. Faulkner and Earl Logan, Jr.
133. *Rapid Prototyping Technology: Selection and Application,* Kenneth G. Cooper
134. *Reciprocating Machinery Dynamics: Design and Analysis*, Abdulla S. Rangwala
135. *Maintenance Excellence: Optimizing Equipment Life-Cycle Decisions,* edited by John D. Campbell and Andrew K. S. Jardine
136. *Practical Guide to Industrial Boiler Systems*, Ralph L. Vandagriff
137. *Lubrication Fundamentals: Second Edition, Revised and Expanded*, D. M. Pirro and A. A. Wessol
138. *Mechanical Life Cycle Handbook: Good Environmental Design and Manufacturing,* edited by Mahendra S. Hundal
139. *Micromachining of Engineering Materials,* edited by Joseph McGeough
140. *Control Strategies for Dynamic Systems: Design and Implementation,* John H. Lumkes, Jr.
141. *Practical Guide to Pressure Vessel Manufacturing,* Sunil Pullarcot
142. *Nondestructive Evaluation: Theory, Techniques, and Applications,* edited by Peter J. Shull
143. *Diesel Engine Engineering: Thermodynamics, Dynamics, Design, and Control,* Andrei Makartchouk
144. *Handbook of Machine Tool Analysis,* Ioan D. Marinescu, Constantin Ispas, and Dan Boboc
145. *Implementing Concurrent Engineering in Small Companies,* Susan Carlson Skalak

Additional Volumes in Preparation

Reliability Verification, Testing, and Analysis in Engineering Design, Gary S. Wasserman

Bearing Design in Machinery, Avraham Harnoy

Mechanical Properties of Engineering Materials, Wole Soboyejo

Industrial Noise Control and Acoustics, Randall F. Barron

Mechanical Reliability Improvement: Probability and Statistics for Experimental Testing, R. E. Little

Industrial Boilers and Heat Recovery Steam Generators: Design, Applications, and Calculations, V. Ganapathy

Mechanical Engineering Software

Spring Design with an IBM PC, Al Dietrich

Mechanical Design Failure Analysis: With Failure Analysis System Software for the IBM PC, David G. Ullman

Implementing Concurrent Engineering in Small Companies

Susan Carlson Skalak
Afton, Virginia

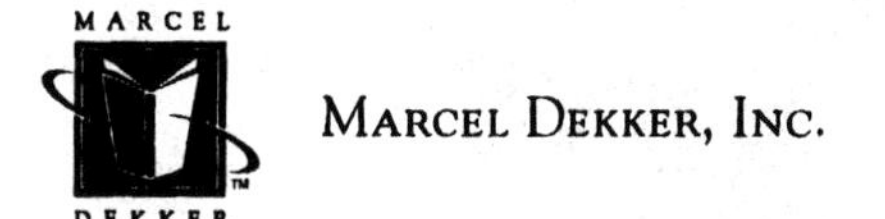

MARCEL DEKKER, INC. NEW YORK • BASEL

ISBN: 0-8247-0762-1

This book is printed on acid-free paper.

Headquarters
Marcel Dekker, Inc.
270 Madison Avenue, New York, NY 10016
tel: 212-696-9000; fax: 212-685-4540

Eastern Hemisphere Distribution
Marcel Dekker AG
Hutgasse 4, Postfach 812, CH-4001 Basel, Switzerland
tel: 41-61-261-8482; fax: 41-61-261-8896

World Wide Web
http://www.dekker.com

The publisher offers discounts on this book when ordered in bulk quantities. For more information, write to Special Sales/Professional Marketing at the headquarters address above.

Current printing (last digit):
10 9 8 7 6 5 4 3 2 1

PRINTED IN THE UNITED STATES OF AMERICA

Preface

The material for this book was developed over a number of years using the insights gained from my industrial and academic experience, my research with graduate students, and our work with small companies. I spent four years as a manufacturing engineer with IBM working with design teams to bring their designs to the manufacturing line. Both product lines on which I worked used concurrent engineering as an overriding philosophy. As a manufacturing engineer, I designed manufacturing stations and tooling to produce the products. In addition, I had sign-off responsibility for the parts of the design that were assembled in my manufacturing stations. With this responsibility came the opportunity to affect the design in the early stages to ensure that it could be easily assembled, in some cases by robots and in others by human operators. Because this design and manufacturing group at IBM functioned as a wholly independent business unit, I was able to perceive and optimize the value added through concurrent engineering and design for small business units. It was a powerful way to improve designs, eliminate assembly errors, reduce design changes, and generally enhance team morale and productivity.

I next spent three years at Georgia Tech to obtain a Ph.D. focusing on design methodology. After assuming an academic position as an Assistant Professor and later as a tenured Associate Professor of Mechanical Engineering at the University of Virginia, I used both my education and work experience to examine design practices in small- to medium-sized companies over a period of seven years. My graduate students and I found that these companies as a rule did not practice concurrent engineering, and tended to pass their designs from department to department without much interaction between the engineers. We were presented with the opportunity to apply the philosophy of concurrent engineering and develop a methodology that could be used by small companies. Our goal was to develop a methodology that could be easily understood and adopted within a company without the use of expensive consultants, which most small companies cannot afford to hire. This book describes this methodology in detail and is a culmination of fourteen years of research and practice with small companies, small business units within large corporations, and student organizations. My chief aim in making a personal career change back to the private sector is to bring this accumulated experience in what really works to small businesses that would benefit from that knowledge. Small business fires the ability of democratic nations to remain creative, innovative, and stable. In a small way, I hope this book helps to fuel that fire.

The book is divided into nine chapters. The first chapter builds a case for adopting concurrent engineering and design within a small company. The second chapter discusses the infrastructure that needs to be in place to implement a successful change to concurrent engineering. Chapters 3, 4, 5, and 6 explain the core of the methodology, for both product design and for manufacturing line design. Chapters 7 and 8 discuss ways in which the methodology can be adapted to include specific company requirements such as environmental or risk reduction techniques. In each chapter, the main ideas are accompanied by examples of designs developed in various industries.

The ideas presented in this book are straightforward and easy to understand, and are easily adaptable to most industries. However, the implementation of these ideas can be very challenging. Implementation of concurrent engineering will require a philosophical change from top management through the ranks of the company all the way to the manufacturing floor. Because most people resist change, the implementation of this methodology requires a strong, public commitment from upper management. Furthermore, the philosophy and methodology described in this book also require commitment to continuous improvement. A dedication to the continuous improvement of the application of the methodology will produce a competitive company that is agile enough to respond to changing market needs.

ACKNOWLEDGMENTS

I would like to recognize the efforts of my graduate students at the University of Virginia. These students worked as a team with me, with each other, with student design teams, and with small companies and independent business units within large corporations to develop the core of this methodology. Natasha Ter-Minassian laid the groundwork for this methodology through her study of design practices in companies throughout Virginia. The other students are all noted as collaborators in the chapters in which their work has relevance, but I would like to mention them again. Hans-Peter Kemser worked on defining the basis for the design methodology, and has since gone on to apply his knowledge at BMW in Germany. Johanne Phair worked on the sister methodology for the design of manufacturing lines. Kevin Allen developed a methodology for assessing and improving the modularity of products, and is now applying his knowledge as a design engineer for the US Navy. Melanie Born developed the risk assessment methods, and is now working as a consultant. Chris Henson developed the basic structure of the outputs for each of the product design steps, and is now a sales engineer in the automotive industry. Lastly, Mark Sondeen spent many hours understanding the needs of small companies from an environmental perspective. He developed design for environment tools and adapted others for use in small companies. He is now applying his knowledge in the aerospace industry. Without these students, this concurrent engineering methodology would have taken many more years to develop, and would look very different from the final form presented. It was a pleasure to work with each of these talented people, who all brought many good and new ideas to the methodology.

I would also like to thank all the companies that participated in our studies. Comdial was most helpful in sharing their time and ideas, and allowing us to be part of their design teams for over a year. I would also like to mention IBM, my first employer, which introduced me to the ideas of concurrent engineering and design.

I would like to thank the National Science Foundation, which supported my work over several summers and supported some of the students mentioned above through the Early Career Development Award. Without their support this work would not have been possible. Lucent Technology and AT&T supported the work described in Chapter 8 through their Industrial Ecology Fellowship.

I would also like to thank the editors at Marcel Dekker, Inc., especially John Corrigan, who first approached me with an interest in this work and patiently waited while we developed the methodology, always reminding me that he was interested in this book, and also Helen Paisner who saw this work to completion. I would like to thank my husband, Tom, whose encouragement and support in this work allowed its completion. Finally, I would like to acknowledge my son, Scott, who sits in my lap as I complete this Preface, and who had the good timing to arrive just as I finished the manuscript.

Susan Carlson Skalak

Contents

1

An Introduction to Concurrent Engineering

In collaboration with Hans-Peter Kemser

Managing a small company has always been a challenge, but today with global markets, the internet, and overnight delivery services, meeting customer needs and delivering quality products on time in a highly competitive market is especially difficult. The temptation for a small- or medium-sized company that has a large market share in its particular niche is to continue doing things the same way it has always done them. Why change when sales are good, the customers are relatively happy with their products, and a product eventually does get out the door even if the schedule slips a little? However, the market is always changing; eventually your company's products will be challenged by domestic or overseas competitors. This book is intended to help change the way small- and medium-sized companies design and manufacture their products to keep ahead of the competition. What manager doesn't want to remain competitive by reducing their time-to-market, improving product quality, and reducing costs? This book introduces a product and process development methodology that can help your company achieve these objectives. Conceptually, the changes needed to achieve more efficient product development are easy, but implementing them is a challenge because it is human nature to resist change unless faced by adversity. Implementation cannot occur overnight, but may take months, even years using continual process improvement as the model for change. This redirection requires a commitment from every employee in the company, from CEO to maintenance worker. The goal of this book is to provide a roadmap that will guide a company to better and more efficient product development and manufacture, and that can be used by every person involved in product development, including design, manufacturing, sales, and service.

This book is intended for small companies, but what is a small company? There are many ways to define a company's size such as by the number of employees or by its sales volume. In this book, we rely on number of employees and number of engineers to define size. Our methodology has been developed for companies having five or more engineers. If a company has fewer than five, then the engineers are usually responsible for seeing a product through the entire design and production cycle. With more than five engineers, the company is usually segmented into functional units, and the design is passed from one function to another. Theoretically, our approach can be used in large corporations; however, most large corporations have their own internally-defined design processes and they would be unlikely to use our approach. Furthermore, our methodology is intended for the products designed by small- to medium-sized companies, which are generally less complex than those of large corporations. For example, the method works well for consumer goods such as toasters and telephones, and for products such as hydraulic cylinders, pumps, and motors. However, the methodology is not intended for complex products such as automobiles and commercial jet aircraft such as the Boeing 777. Finally, the methodology is intended to help companies that are primarily situated at one location, and design and manufacture the majority of the new product in-house.

1.1 IS THIS YOUR COMPANY?

Although few studies have been conducted in small- and medium-sized companies, those studies found that small companies operate similarly [Albin and Crefeld, 1994; Radcliffe and Harrison, 1994; Carlson, *et al.*, 1997]. These small companies are characterized by many of the following attributes:

- Limited resources,
- Engineers' time spread across multiple projects,
- Informal communication among personnel and often little or no input from manufacturing into the development process,
- No sense of ownership,
- Design methods governed by rules of thumb,
- Little documentation of designs and of lessons learned,
- Few formal project management and planning skills,
- Low priority of trial runs and prototyping,
- Concentration on short term goals, and
- Minimal stand-alone influence on industry as a whole.

Furthermore, most of the companies studied practice a sequential engineering design process as shown in Figure 1.1. In a sequential engineering (SE) process, sales and marketing define the parameters of the new product, including the sales

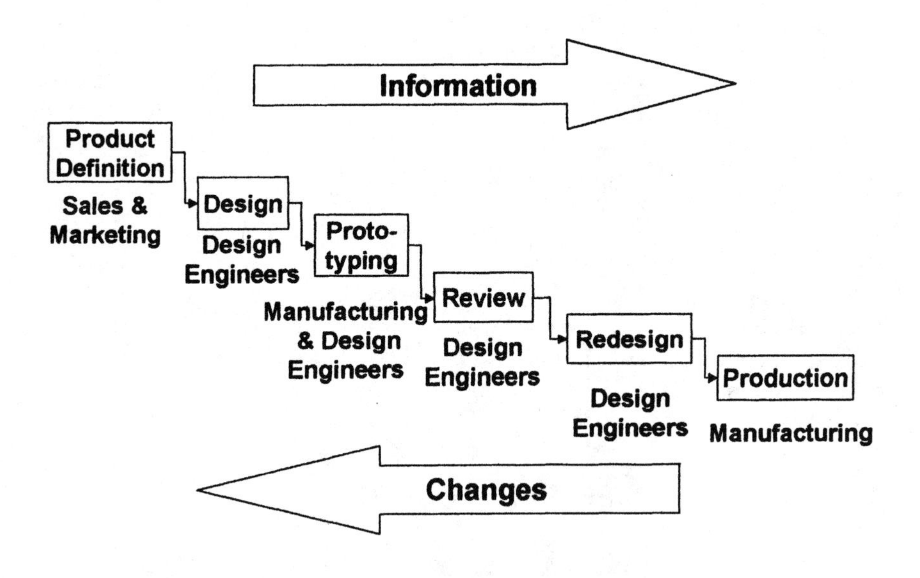

Figure 1.1 Flow diagram of a sequential engineering process.

price, and requirements, which are based on customer needs. Then, they present a product definition document to the design engineers. Using this information, the engineers then define the product specifications, design the new product, and pass detail design drawings to manufacturing. If the requirements ask for a prototype, and the product is not too far behind schedule, then manufacturing will build a prototype. The results are then reviewed by the design engineers. Problems that are identified in the prototype, and any difficulties that manufacturing finds in producing the product as designed are discussed. The design engineers then make changes to the product to accommodate manufacturing and any problems found in the prototype. Finally, production will occur after all problems are solved or minimized to an acceptable level. In this model, the groups rarely interact beyond the hand-off of the design from one stage to the next. Most of the design changes occur in the redesign phase of the sequential engineering process, causing additional costs and time delays. Since the next development phase can only begin after the preceding one is completed, the product launch date is delayed with each change.

Another complicating factor in the development of new products within smaller companies is the lack of focus on the future [Radcliffe and Harrison,1994; Braiden, Alderman and Thwaites, 1993]. Small companies focus on current products, fixing problems that arise in the field with the products already on the market, and they spend little time on designing new products for customer demands. This leads to a "fire-fighter" mentality in the engineers, in which they focus on answering problems from the field, and new product development takes low priority. When viewed from a distance, this problem is easy to detect and intellectually, it seems clear that this strategy will result in long term decline of customer satisfaction with the current products. The company's products will eventually become outdated as the engineers scramble to keep up only with today's demands. The company feels it is very responsive to customer satisfaction, but what they don't see is that eventually they will not be able to keep up with new demands as the market passes them by and competitors deliver more up-to-date products.

How can a company escape from this destructive spiral? They must fix problems up-front in the design of the product, not when it is already in production. They must change their operations to become more concurrent and team oriented.

1.2 WHAT IS CONCURRENT ENGINEERING?

Concurrent engineering (CE) has been described by many different people and organizations. Companies often practice some form of CE or portions of it without defining it as concurrent engineering. There are many other names that have been applied to the same principles such as simultaneous engineering, integrated product and process design, concurrent design, etc., and these terms are still used. Concurrent engineering is defined by different people as different things. For example, the original definition that is frequently referenced was developed in 1986 by the Institute for Defense Analysis:

> *Concurrent engineering is a systematic approach to the integrated, concurrent design of products and their related processes, including manufacture and support. This approach is intended to cause the developers, from the outset, to consider all elements of the product life cycle from conception through disposal, including quality, cost, schedule and user requirements.* [by permission from the Institute of Defense Analysis, Report R-338]

Many other definitions have been published since this one. Most focus on integrating and managing the design process to result in a shortened time-to-market. The following list indicates how differently authors view concurrent engineering.

- *A systematic approach to integrated product development that emphasizes the response to customer expectations. It embodies team values of cooperation, trust, and sharing in such a manner that decision making proceeds with large intervals of parallel working by all life-cycle perspectives early in the process, synchronized by comparatively brief exchanges to produce consensus* [by permission from Cleetus, "Definition of Concurrent Engineering," *CERC Technical Report Series, Re-*

search Notes, CERC-TR-92-003, West Virginia University, Morgantown, W.V., 1992].

- *CE is a management and engineering philosophy for improving quality and reducing costs and lead time from product conception to product development for new products and product modifications* [Creese, R. and L. Moore, "Cost Modeling for Concurrent Engineering," Cost Engineering, Vol. 1, 1990, pp. 113-124. By permission of AACE International, 209 Praire Ave, Suite 100, Morgantown WV 25601 USA, Phone 800-858-COST/304-296-8444. Fax 304-291-5728, Internet http://www.acei.org Email: info@aacei.org].

Because these definitions cover such a wide range of concepts from team empowerment to cost reduction, the following attributes have been gathered to characterize a CE design process:

- Customer focus and involvement,
- Early and continual involvement of suppliers in the design process,
- Cross-functional, self-directed, empowered teams,
- Incremental sharing and use of information,
- Life-cycle focus,
- Systematic and integrated approach,
- Concurrent (parallel) design teams,
- Early use of Design for X (DFX) tools,
- Use of modern tools such as CAE, CAD, CAM, finite element analysis, etc., and
- Continuous improvement of all processes.

How does CE differ from SE? The number of design changes in CE are minimized in the downstream stages of design, because every person who has a stake in the product's life cycle is involved in the design process from the beginning. Therefore, issues such as maintenance, manufacturing, and customer use, are addressed from the beginning of the process by the cross-functional design team. Since a cross-functional team is used and customers and suppliers are involved in the process from product definition, the entire development time is reduced. A representation of typical time savings of using CE is shown in comparison with SE (Figure 1.2). When implementing CE, one caution that should be heeded is that the front-end phases become much longer in comparison with SE. Notice that only 3% of the development process was spent on planning in SE, and 27% was spent on design. Compare that with the time used planning in the CE process. The time required was more than 10 times greater, the planning process took about one-third of the entire process time, and design 22%. In SE, only one-third of the development time was spent on planning and design, and the rest of the time was spent fixing problems that arose when the product was transferred to the manufacturing organization. In contrast, over half of the development

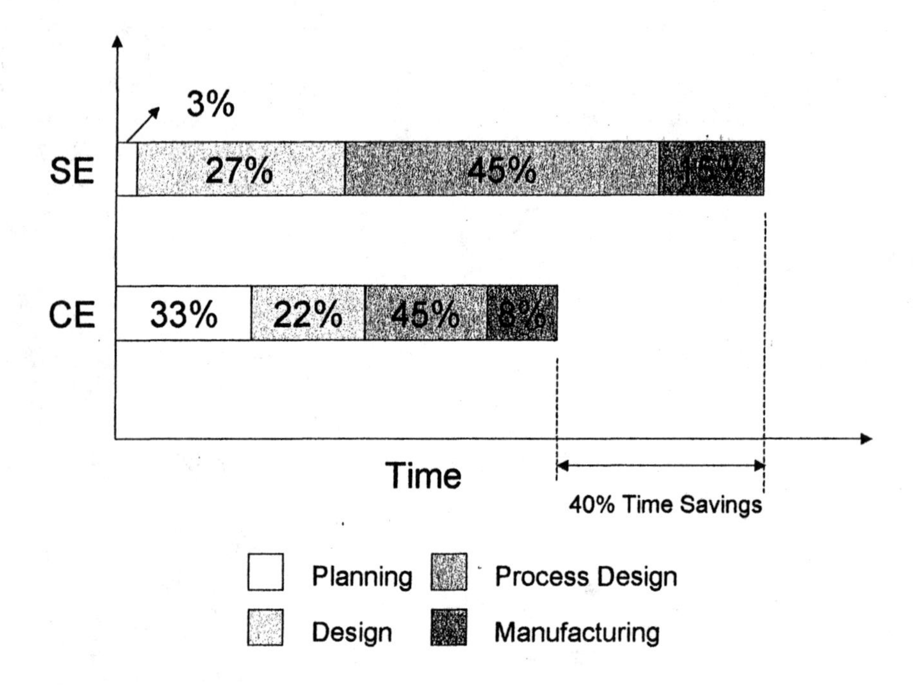

Figure 1.2 Comparison of SE with CE, showing time savings with CE use. [US Air Force R&M 2000 Process Study Report, 1987.]

effort was spent on planning and design in CE, yet there was a 40% savings in time using CE. The problems in the CE process were fixed up-front in the cross-functional design teams, so that when the product got to the manufacturing organization, few problems had to be fixed, and fewer still once the product was in production. Organizations that choose to switch to CE must be aware of the necessary requirement of time in planning and design. Managers should not benchmark a CE process with an SE process, and expect the duration of the stages to coincide. The time savings come at the end of the design process, not in the early stages.

1.3 THE BENEFITS OF CONCURRENT ENGINEERING

Many large corporations have used concurrent engineering since the early 1980's. Therefore, there has been considerable data collected on the benefits of the use of CE in these organizations. The following list of benefits were achieved through the successful implementation of CE in large corporations [Linton, *et al.*, 1991, used with permission of the National Security Industry Association].

Development and production lead time reductions:

- Product development time reduced by up to 60%
- Production spans reduced 10%
- Total microelectronics fabricating process time reduced up to 46%.

Measurable quality improvement:

- Yield improvements of up to 4 times
- Field failure rates reduced up to 83%.

Engineering process improvements:

- Engineering changes per drawing reduced up to 15 times
- Early production engineering changes reduced by 50%
- Scrap and rework reduced up to 87%.

One of the most cited benefits of using CE is the reduction in time-to-market, which is the time required from product definition to production of the first product. Table 1.1 shows the reductions of development times for fast innovators. Table 1.2 shows the reduced time between ordering and shipping. Note that the major contributing factor for time reduction in these examples was the use of a multidisciplinary team.

An example of a medium-sized organization (750 employees, over 60 engineers) that has successfully implemented concurrent engineering is Comdial Corporation, located in Charlottesville, Virginia. Comdial is a world leader in the design and manufacture of business telephone systems, and despite its size, it has competed successfully with giants such as AT&T and Northern Telecom. Comdial began the journey to implementing concurrent engineering in 1984. Because of its policy of continual assessment and improvement, their concurrent engineering process today is much more integrated and effective that it was in 1984. Comdial was faced with a crisis in 1984: they had entered the retail phone market with the divestiture of AT&T. However, they were unable to compete effectively in this market with inexpensive phones that were flooding in from off-shore competitors. In 1984, they decided to focus on business telephone systems only and had to deliver a new product to the fall tradeshow, giving them only 6 months to develop a product that normally took them 2-3 years. With the guidance of their vice-president, who was well-versed in teaming and Japanese development practices, Comdial made the change to concurrent engineering. Tables 1.3 and 1.4 show the improvements Comdial has made since 1984 in its key measurements of time-to-market and quality.

Table 1.1 Time-to-market comparisons for large corporations that have been classified as fast innovators [This figure is reproduced from Biren Prasad's *Concurrent Engineering Fundamentals: Integrated Product and Process Organizations* Figure 4.19a, page 210 in Vol. 1, ISBN 0-13-147463-4, published by Prentice Hall, 1997. Permission received from Dr. Biren Prasad, Author May 2002].

Fast Innovators					Major Contributing Factors
Company	Product	Best Development Time			
		Before CE (mos.)	After CE (mos.)	Reduction	
ABB	Switching systems	48	10	79%	CMS
AT&T	Phones	24	12	50%	ACM
British Aerospace	Airplanes	36	18	50%	MS
Digital Equipment	PCs	30	12	60%	ACMS
Ford	Cars	60	42	30%	---
GM	Engines	84	48	43%	MS
GM/Buick	Cars	60	41	32%	MS
Goldstar	Telephone systems	18	9	50%	CM
Honeywell	Thermostats	48	12	75%	MS
Honda	Cars	60	36	40%	---
Hewlett-Packard	Printers	54	22	59%	ACMS
IBM	---	48-50	12-15	70-75%	ACM
Motorola	Mobile phones	36	7	81%	ACM
Navistar	Trucks	60	30	50%	MS
Warner Electric	Clutches/ Brakes	36	9	75%	ACM
Xerox	Copiers	60	24	60%	ACM
Legend: A: Analytical methods and tools, M: Multi-disciplinary team C: Computer integrations, S: Suppliers on the project team					

Table 1.2. Order-to-ship improvements for large, corporate, fast producers [This figure is reproduced from Biren Prasad's *Concurrent Engineering Fundamentals: Integrated Product and Process Organizations* Figure 4.19b, page 210 in Vol. 1, ISBN 0-13-147463-4, published by Prentice Hall, 1997. Permission received from Dr. Biren Prasad, Author May 2002].

Fast Producers			
Company	Product	Order-to-Ship	
		Before CE	After CE
Brunswick	Fishing reels	3 wks	5 days
GE	Circuit breakers	3 wks	3 days
HP	Test equipment	4 wks	5 days
Motorola	Pagers	3 wks	¼ days

Table 1.3 Comdial's time-to-market improvements [Used with permission from Comdial Corporation, 106 Cattlemen Road, Sarasota, FL 34232].

Product	Time-to-Market		
	1984	1998	Improvement
Printed circuit board design	2 mo	1.5 mo	25%
Telephone systems	18 mo	9 mo	50%
Key service unit	24 mo	12 mo	50%

Table 1.4 Comdial's quality improvements [Used with permission from Comdial Corporation, 106 Cattlemen Road, Sarasota, FL 34232].

Measure	1984	1998	Improvement
Rate of field returns	5%	1%	4%
First pass success rate	90%	98%	8%

In general, the following benefits can be achieved by implementing CE within an organization:

- Improvement in product quality,
- Reduction of product cost and lead time,
- Lower test and inspection costs,
- Reduction in number of engineering changes,
- Increase in profitability,
- Improvement in competitiveness,
- Increase of ownership among employees, and
- Higher customer satisfaction.

These improvements can only occur when all aspects of CE are implemented. Mixing CE with SE processes leads to chaos and employee frustration. CE can only be successfully implemented with the commitment and dedication of senior management.

1.4 HOW CONCURRENT IS YOUR COMPANY NOW?

The following questionnaire is included to help you look at the current product development practices of your company and to identify areas that need improvement. The questionnaire [*Concurrent Engineering* by Carter, Baker, © Adapted by permission of Pearson Education, Inc., Upper Saddle River, NJ].is divided into four parts: Organization, Communication Infrastructure, Requirements, and Product Development. After completing a CE implementation, an organization should be able to answer all these questions positively.

Organization

- Are the specifications and priorities for the assigned tasks understood by each team member?
- Is the product development process understood by each team member of every team?
- Are design decisions made by a cross-functional team?
- Do customers and vendors participate in your design decisions?
- Is a cross-functional team (including customers and vendors) responsible for the development of engineering specifications, scheduling, and system design specifications?
- Are teams rewarded for their contributions?
- Is adequate training provided for each member of the team regarding procedures, tools, and standards they should know?
- Are cross-functional teams provided adequate team training?

Communication Infrastructure

- Are the tools for each team member integrated within the company?

- Are electronic mail capabilities available to each team member?
- Are query and online reporting capabilities available to each individual?
- Are interactive product data browsers available to each individual?
- Are technical reviews conducted at appropriate milestones?
- Are the management team and the customer concurrently informed of problems and their status?
- Do individuals and teams have electronic access to company-wide product development data that includes data from customers and vendors?
- Is the product development data stored, controlled, changed, and versioned in a common computer database?
- Is the data in the product development database interoperable among the various design automation tools?
- Are evolving product requirements, specifications, and development data under automatic changes and versioning controls?
- Are action items, problem reports, enhancement requests, and all other decisions analyzed to continuously improve the product development process?

Requirements

- Does your company measure best-value product designs for cost, functionality, fitness for use, reliability, performance, and supportability?
- Does your company use design standards to ensure product testability, manufacturability, supportability, etc., and are these standards regularly reviewed and improved?
- Can the cross-functional team access the product life-cycle specifications as part of decision support?
- Are adequate planning methods used for deciding upon product requirements?
- Are engineering and process specifications validated to the customer needs?

Product Development

- Is the component library system linked to a decision support tool to assist each designer in making component or unit selections?
- Are analysis tools used during conceptual and detail design?
- Are goals for product and process improvements in place?
- Are major project decisions and the factors leading to them documented, distributed, and analyzed for guidance on other projects?
- Are product designs, development processes, specifications, and tools concurrently analyzed and continuously improved as a part of a company-wide optimization strategy?

1.5 REQUIREMENTS FOR SUCCESS?

Taking the tremendous step to move from an *ad hoc* design process to the more systematic concurrent engineering process requires rethinking how processes are performed. Organizations must change their structure, product and process development approach, interaction between employees, and their interaction between the customer and supplier. New methods must be embraced and adopted such as employee empowerment, total quality management, and continuous improvement and innovation. These methods lead to a change of culture of the entire organization and can only be achieved with the full commitment of senior management.

Since the benefits of CE are so obvious and dramatic, and so many organizations have already implemented CE successfully, the question is why don't all companies change? One reason is given below.

> *One of the difficulties in bringing about changes in an organization is that you must do so through the persons who have been most successful in that organization, no matter how faulty the system or organization is. To such persons, you see, it is the best of all possible organizations, because look who was selected by it and look who succeeded most within it. Yet these are the very people through whom we must bring about improvements.*
>
> *-George Washington, Second Inaugural Address*

The successful implementation of CE can only be achieved if senior management embraces and supports the change by establishing clear guidelines of how the change will take place, and how teams and individuals will be awarded in the adoption of the new system. Management must lead employees through the operational changes by explaining why the changes are necessary and what may happen if changes do not occur.

1.6 PHILOSOPHY OF THIS BOOK

This book was designed to help small- to medium-sized firms adopt concurrent engineering as a design management methodology within their company. Concurrent engineering has been around for more than a decade, and many large corporations have found that its use has resulted in significant gains in productivity accompanied by a reduction in time-to-market, costs, and quality problems. Smaller corporations will be able to see similar results with the use of concurrent engineering. This book is intended as a "how-to" guide for managers and engineers to help them learn a systematic approach to concurrent engineering, and one especially developed for use in small companies. The engineering design process is not always well understood, and this book will guide the practicing engineer and manager through the process step by step, using industrial examples throughout.

The book begins with an overview of concurrent engineering, its challenges, and its benefits. Next, teams and how they are used in the context of concurrent engineering are discussed. Then, a roadmap is provided that guides the engineer/manager through the five phases of design: planning, conceptual design, design, production preparation, and production/service. Each of the phases is divided into design steps accompanied by a detailed explanation of each step, the team members needed to perform the step as well as the support members, and the output of each step. What is truly unique in our approach to concurrent engineering is that we have three main design models, one for original design (a new design), one for evolutionary design (a major update of an existing product), and another for incremental design (a minor product change). These models can be tailored to fit the unique characteristics of a company as well as any design project. In addition to addressing product design, we also have developed a methodology for the design of manufacturing lines to produce the product, and have also developed tools and techniques to help companies address issues in designing their products to be more environmentally friendly. This book will lead you through a step-by-step process to implement concurrent engineering in your company. We hope you will find this book useful in improving your company's performance.

REFERENCES AND BIBLIOGRAPHY

Albin, S. and P. Crefeld, "Getting Started: Concurrent Engineering for a Medium-Sized Manufacturer," *Journal of Manufacturing Systems*, Vol. 13, No. 1, pp. 48-58, 1994.

Braiden, P., N. Alderman and A. Thwaites, "Engineering Design and Product Development and its Relationship to Manufacturing: A Programme Case Study Research in British Companies," *International Journal of Production Economics*, Vol. 30-31, pp. 265-272, 1993.

Carlson-Skalak, S., H. Kemser, and N. Ter-Minassian, "Defining a Product Development Methodology With Concurrent Engineering for Small Manufacturing Companies," *Journal of Engineering Design*, Vol. 8, No. 4, December 1997, pp. 305-328.

Carter, D. and B. Baker, *Concurrent Engineering: The Product Development Environment for the 1990's*, Addison-Wesley, New York, 1992.

Cleetus, K., "Definition of Concurrent Engineering," *CERC Technical Report Series, Research Notes*, CERC-TR-92-003, West Virginia University, Morgantown, W.V., 1992.

Creese, R. and L. Moore, "Cost Modeling for Concurrent Engineering," Cost Engineering, Vol. 32, No. 6, 1990, pp. 113-124.

Institute for Defense Analysis [IDA], Report R-338.

Linton, L., D. Hall, K. Hutchinson, D. Hoffman, S. Evancyuk, and P. Sullivan, "First Principles of Concurrent Engineering: A Competitive Strategy for Product Development," *CALS/CE Electronic Systems Working Group Report*, National Security Industry Association, Washington, D.C., 1991.

Prasad, B., *Concurrent Engineering Fundamentals: Integrated Product and Process Organizations*, Vol. I, Prentice Hall PTR, Upper Saddle River, N.J., 1996.

Radcliffe, D. and P. Harrison, "Transforming Design Practice in a Small Manufacturing Enterprise," *Proceedings of the 1994 ASME Design Technical Conference, Design Theory and Methodology*, pp. 133-140, 1994.

U.S. Air Force R&M 2000 Process Study Report, US Air Force, Oct. 1987.

2

Succeeding with Concurrent Engineering

In collaboration with Hans-Peter Kemser

Changing from a sequential design organization to a concurrent organization requires changes in organizational structure, personnel management, and business practices. Organizations must change the way they think about their product and process development approaches, employee interactions, and customer and supplier relationships. A successful implementation of concurrent engineering involves rethinking how processes were executed in the past, and moving toward employee empowerment and open communication between all organizations. All these changes require a modification in the organization's culture that can only be achieved with the support and commitment of senior management. In this chapter, we will look at organizational structures and how they can contribute to the success of concurrent engineering, teams and the role they play in concurrent engineering, and what barriers the company may face as they embrace these changes.

2.1 ORGANIZATIONAL CHANGE AND CE

There are three components that must be examined when considering a change in organizational structure: the organizational structure itself, personnel practices, and business practices and procedures [Syan, 1994]. Each of these is defined below.

- *Organizational structure*: The organizational structure within a company is usually governed by the product. For example, in Comdial, Incorporated, the company is structured by functional units such as design, manufacturing, and administration (which includes personnel, purchasing, etc.), and within each function the structure is defined by the tasks performed within the function. For example, Comdial designs and manufactures phone systems, and the design function is divided into three main groups: mechanical hardware, electrical hardware, and software. The main issue that must be addressed within a company's organizational structure is that it allows for empowered teams and communication and cooperation among functions.

- *Personnel practices*: Team leaders should report to senior management directly as well as to their functional managers. The existing reward structure should be revised to reward team performance, and the team should be able to award individual performance. The reward could be for finishing a project on time or before the due date, for high quality product designs that result in few quality problems, and for meeting or beating cost targets.
- *Business practices and procedures*: The relationship with vendors and customers will change. Both these groups should play a more active role in the development of new products. Also, processes and procedures that compete with each other should be eliminated.

We will now examine the first two of these components in more detail and discuss options that are available. The last component will be discussed in Chapters 4, 5, and 6.

2.1.1 Product Development Organizational Structures

There are four types of product development structures: functional, lightweight, heavyweight, and autonomous [Wheelwright and Clark, 1995]. Figure 2.1 shows these structures graphically. Each of these structures is discussed in more detail below.

The functional product development structure is often used by small organizations practicing CE. In this type of structure, the organization is divided along disciplinary lines, such as marketing, manufacturing and engineering and is supervised by a functional manager. Projects are divided along functions, and any issues that require coordination across these functions are handled by infrequent meetings. Typically, as each function completes its work, it is passed to the next function in sequential fashion, as indicated by the arrows in Figure 2.1 in the functional structure block. The advantage of this type of structure is that each function can develop their specializations, especially in large organizations with many resources. There are several weaknesses of this organizational structure. The structure can be quite bureaucratic especially in larger companies, and there is limited coordination between structures. Also, product development is often slowed by iterations of the projects between functions.

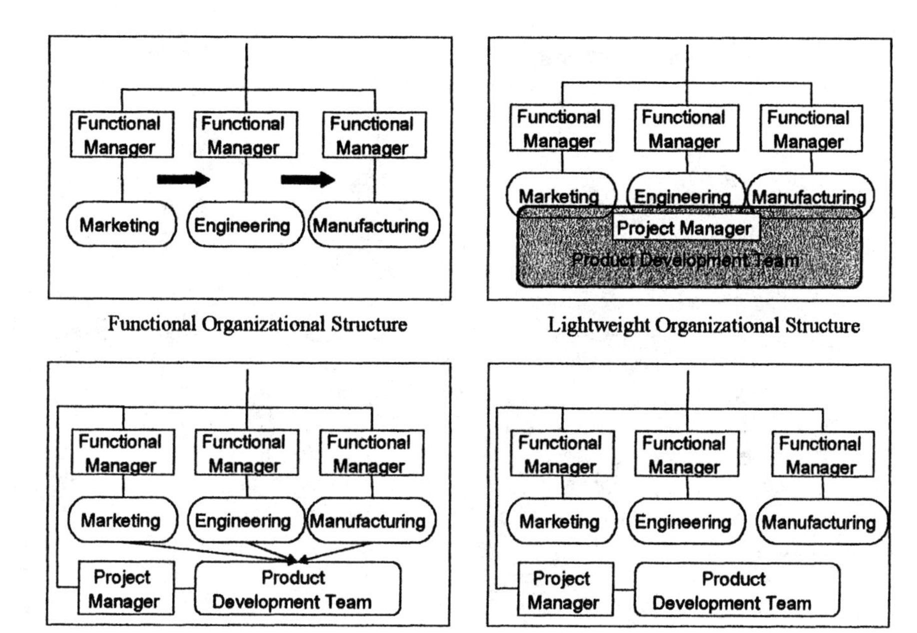

Figure 2.1 Product development organizational structures [Reprinted with the permission of the The Free Press, a Division of Simon & Schuster, Inc., from *Leading Product Development: The Senior Manager's Guide to Creating and Shaping the Enterprise* by Steven C. Wheelwright and Kim B. Clark, Copyright ©1995 by Steven C. Wheelwright and Kim B. Clark.]

In the lightweight structure, all functional areas involved in a product development project have representatives in the development team. The work still remains in the functional areas, however, the projects are scheduled, coordinated, and monitored by a project manager in charge of the team of functional representatives. In this structure, the power still resides with the functional managers with the project manager having little authority. All allocation of resources are made within the functions, and the project manager's task is to keep the functional managers informed of the project's progress and needs.

In the heavyweight structure, team members are selected for the product development team from the functional areas. Although still attached to their functional groups, they are assigned to the product development team for the duration of the project. The project manager is responsible for all project personnel and the success or failure of the project. Much of the work still occurs in the functions, but that work is overseen by the project manager and the project team rather than by the functional manager.

In the autonomous organizational structure, as in the heavyweight and lightweight structures, product development team members are assigned from functional areas. In contrast to the other structures, the project manager has full control over the team members and is responsible for evaluating their performance rather than having that responsibility reside with or shared with the functional managers. The autonomous team is often used as a 'tiger team' or 'skunk works' in which high performing members work independently of the rest of the company to develop innovative products.

There is not one right organizational structure for a company. The structure is often dependent upon the new product to be developed. For example, if the company wants to develop an innovative product that is very different from their existing product lines, then an autonomous structure is recommended if they can spare the resources. Many small companies that practice CE and produce a variety of products use a lightweight structure. The team members in this structure-type report to their functional managers to present the current status of the project. The functional managers can make decisions regarding resource allocation to help keep the projects on time and on budget. However, senior management must ensure that conflicts that arise between project managers and functional managers are resolved quickly to avoid unnecessary delays and favoritism among projects. The lightweight structure is especially well suited for the quick development of follow-on projects of existing products, which are the majority of development projects undertaken by small companies.

2.1.2 Teams and Teamwork

As discussed in the section on organizational structures, teams play an important role in the development of products. An in-depth discussion of the formation, development and management of teams is beyond the scope of this book. However, there is an abundance of literature on the subject as indicated at the end of this chapter. The product development teams in the organizations discussed in the previous section were formed by pulling personnel from different functional areas. However, getting these people to work together to form a team was a challenge. First, let's start by defining a team. Katzenbach [1993] defined a team as:

> *...a small number of people with complementary skills who are committed to a common purpose, set of performance goals, and approach for which they hold themselves mutually accountable.* [Katzenbach, J. and D. Smith, The Wisdom of Teams: Creating the High Performance Organization, Harvard Business School Press, Boston, 1993, reprinted with permission from the Harvard Business School Press.]

The basic elements of a team are shown in Figure 2.2. The sides and the center of the triangle show the elements necessary to obtain the outcomes of products, performance and personal growth.

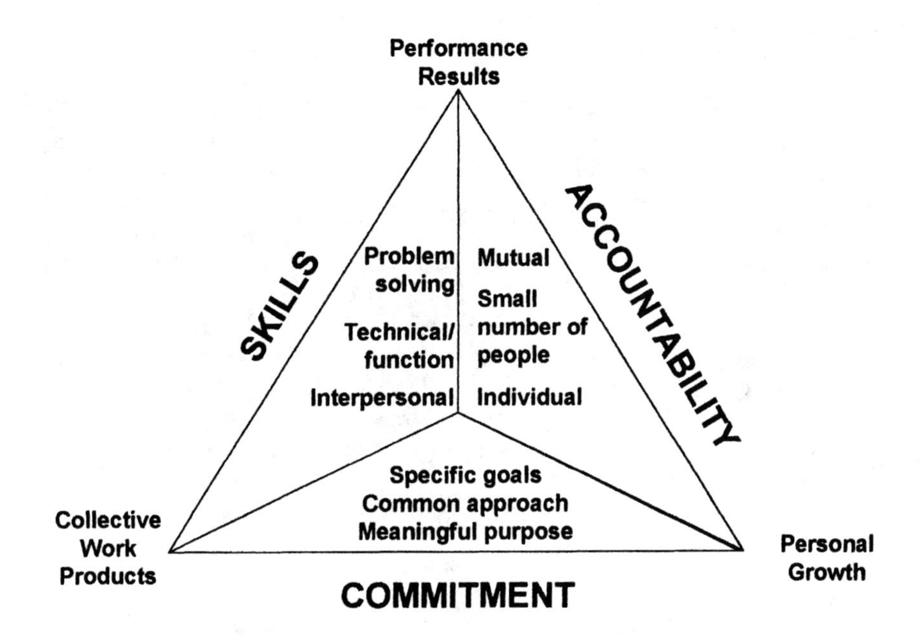

Figure 2.2 The basics of teams that includes their characteristics, their requirements, and their outcomes. [Katzenbach, J. and D. Smith, *The Wisdom of Teams: Creating the High Performance Organization*, Harvard Business School Press, Boston, 1993, reprinted with permission from the Harvard Business School Press].

A key step in the successful use of teams is training. To get a group of people from different functional areas together and have them function as a team can be difficult, especially if cross-functional teams are not a regular part of your business. In these situations, training at the start of implementation is imperative for success. Training can help speed team effectiveness, and the training is most effective when the group is given a real or a simulated problem to solve under the guidance of a trainer. Another effective training method is the use of a facilitator. The facilitator can help keep the team on track and help team members learn to resolve conflicts, which are common in cross-functional teams having members of different educational backgrounds, such as liberal arts, business, or engineering.

In addition to training, there are several other things a team needs to be successful:

- Creativity,
- Expertise,
- Leadership,
- Organization,

- Resources, and a
- Method or strategy for handling conflict.

Creativity and expertise must be carefully managed. The expertise in the teams is provided when members are chosen from functional areas such as marketing, engineering design, manufacturing, and maintenance. However, this expertise must be balanced with creativity. Engineers tend to recommend those solutions and designs with which they are familiar. For example, if you have an engineer with significant expertise in sheet metal fabrication, it is unlikely he will recommend injection-molded parts. Creative solutions can spur the company to new product lines, however, creativity needs to be tempered with reality. For example, the development of new technologies should not be undertaken in product development projects with tight deadlines. Leadership should be able to balance creativity and expertise to the benefit of the team and the company. Leadership can come from many places, such as senior management, project leaders, and team leaders. Team leadership should be shared, and should change often over the course of the project as different expertise is needed. Organization and resources are the purview of the management.

In addition to choosing teams members for their complementary skills, personality plays an important part in team dynamics and success. Members of cross-functional teams play two different roles: 1) they represent their functional area with their technical skills and knowledge, and 2) they play a range of behaviors based on personality [Stetter and Ullman, 1996]. Table 2.1 lists nine different team roles that should be represented to ensure a high performance team. It should be noted that some members can play multiple roles.

One of the primary hurdles to the successful implementation of teams is empowering the team to make decisions. Managers must learn to relinquish many of the decisions that they once controlled. Senior management's role has a more global flavor than in the past. These roles include:

- Definition of mission and boundaries,
- Assurance of strategic fit,
- Provision of career opportunities and growth,
- Support,
- Provision of resources, and
- Measurement of progress through review.

Management must ensure that the project undertaken fits with company strategy, and then define the boundaries and mission of the project team to ensure they do not go off in directions that do not fit with company strategy. Managers should select team members not only for their contribution to the team, but also to provide new opportunities and growth for their employees. Management is also responsible for ensuring that the team has the support and resources that it needs to

Table 2.1 Team roles based on behaviors and personality [Stetter, R. and D. Ullman, "Team Roles in Mechanical Design," *Proceedings of the 1996 ASME Design Engineering Technical Conference and Computer in Engineering Conference*, 96-DETC/DTM-1508, August 18-22, 1996, Irvine, CA. Reproduced with permission from ASME.]

Team Role	Definition
Organizer	A reliable person concerned about the practical aspects of the design process.
Motivator	A confident person in charge of the schedule and goals of the team.
Pusher	A dynamic person forcing a design team to work faster.
Solver	A creative person predominately generating solutions.
Gatherer	An extroverted person searching for information and communicating with others outside the team.
Listener	A perceptive person perceiving and combining the ideas and statements of others.
Completer	A conscientious person eliminating the last flaws of the design.
Specialist	A dedicated person with extensive knowledge in a specialized field.
Evaluator	A strategically thinking person concerned about alternative solutions.

complete the design on time and within budget. Finally, senior management tracks the progress of the team through reviews at key points throughout the life of the project. Notice that managers are no longer asked to provide a schedule or key technical decisions. Those are now made by the team.

Empowerment can enhance a company's performance. Research of the 25 finalists of 1995's "America's Best Plants" showed one common characteristic: employee empowerment [Sheridan, 1996]. Management should remember that empowerment means that decisions should be made at the point of maximum information, which is among the employees, rather than at the management level.

> *Empowerment has two parts: a management part and a psychological part. The management part involves sharing power (e.g., pushing decisions down into the organizations as far as possible, sharing information widely, employee participation in goal setting and problem solving, reducing rules and restrictions). The psychological part means helping people achieve a*

*powerlessness, people seeing the impact of their actions).[*Coger, J. and R. Kanungo, "The Empowerment Process: Integrating Theory and Practice", *Academy of Management Review,* Vol. 13, pp. 471-482, 1988. Quoted with permission from the Academy of Management Review.*]*

Team empowerment is a difficult change to employ in a company that has been operating under a more autocratic system. However, once the change is made, both manager and employees can be come more fulfilled. Managers are no longer asked to make detailed decisions about the design for which they have little information. Rather, they can concentrate on more global issues such as strategy and resourcing. Employees are more fulfilled because their decisions have direct impact on the product, and they feel they are now able to make real contributions to the company.

2.1.3 Do You Have A Vision?

Every successful organization needs a clear vision, especially when undertaking a major change of operation such as the implementation of CE. Senior management must develop a clear picture of the company's future directions and long term goals that can be understood by every employee. This vision is particularly key to the successful use of teams. Vision statements can be used at the corporate, division and project levels. No matter at what level the statement is used, it should have the following characteristics [Kotter, 1995]:

- Picture of the future,
- Easy to communicate,
- Appeals to employees, customers, and stockholders,
- Goes beyond numbers that are typical in five-year plans,
- Helps to clarify the direction in which the organization needs to move.

In addition to a vision statement, a mission statement and objectives needs to be developed. To clarify the difference between these terms, the following quotes should be helpful.

A vision differs from a mission statement in that a mission statement is a statement of what business we are in and sometimes our ranking in that business. The mission statement names the game we are going to play...A vision is more philosophy than how we are going to manage the business [Block, P., "The Empowered Manager," Positive Political Skills at Work, Jossey-Bass, 1987, p. 115. Copyright © 1987 Jossey-Bass. Reprinted by permission of Jossey-Bass, Inc., a publishing unit of John Wiley & Sons, Inc.].

You should be able to communicate the vision within five minutes or less and receive a reaction that indicates both understanding and interest [Kotter, J., "Leading Change: Why Transformation Efforts Fail," Harvard

Business Review, March-April 1995, pp. 59-63. Quoted with permission from Harvard Business School Press].

A vision is very different from ...goals and objectives. Goals and objectives are basically a prediction of what is to come. Predictions of what we are going to do in the next week or month or quarter are basically an extension of what we have done the last week month or quarter [Block, P., "The Empowered Manager," Positive Political Skills at Work, Jossey-Bass, 1987, p. 115. Copyright © 1987 Jossey-Bass. Reprinted by permission of Jossey-Bass, Inc., a publishing unit of John Wiley & Sons, Inc.].

Developing a vision statement takes time and often requires many iterations. Keep the statement clear and easy to understand. To help you draft your own vision and mission statements, some example statements from small, successful companies are given below.

- ATK/Bay Area Incorporated
  Vision Statement: *Our vision is to be recognized as the preferred provider of remanufactured engines and related items; providing a standard of quality and service that exceeds expectations and by which all others are measured.*
  Mission Statement: *ATK/Bay Area Inc. is a truly unique company. We provide an environment where people want to come to work because they can grow both personally and professionally, share in the success of the company and enhance their quality of life. We constantly strive to provide remanufactured engines and related items of a consistently high quality at a market price that reflects a competitive cost for the level of quality and service provided. Mutually beneficial long-term relationships with customers, suppliers and employees is of key importance to us. We constantly look for better ways to support them and will respond to their changing requirements by broadening our capabilities and investing in new technologies. We believe in always doing what we say. If we make a mistake we will admit it, learn from it and carry on. The achievement of our mission will require all involved to constantly strive for improvement in quality, service, cooperation, communication, mutual respect and trust.* [Reprinted with permission from ATK/Bay Area Incorporated, 2000]

- Ionics, Incorporated
  Mission Statement: *To be a leading provider of clean and purified water as well as a supplier of equipment and systems for water and wastewater treatment worldwide. This will be accomplished by:*
  - *Anticipating and meeting the water quality needs of people, industry and our environment,*

- *Utilizing proprietary separations technologies and innovative marketing, distribution and financing methods to provide customers with products and services which deliver superior value.*

Preservation and enhancement of our environment and quality of life is the ultimate legacy one generation can bequeath to future generations. [Quoted with permission from Arthur L. Goldstein, Chairman and CEO of Ionics, Inc. 2000]

2.2 BARRIERS TO CHANGE

Before the implementation of CE can begin, there are many barriers to change that must be overcome. In general, there are two types of barriers: technical and organizational. The technical barriers are, in general, communications related. The biggest technical barrier is usually the lack of a communication system. Employ ees should be able to communicate and send information easily, which leads to the second technical barrier. Often, the company lacks a common database in which information can be stored and retrieved easily by different departments. Often in companies, each function has its own information system that cannot be accessed by other functions. Finally, the lack of a CAD/CAM system can be a hindrance, not only internally, but also in the communication with vendors and customers. The technical barriers are the easiest to address. However, the organizational barriers pose a bigger challenge.

There are many organizational barriers that can prevent the successful adoption of CE. All these barriers must be addressed by senior management. Without the support of senior management, a successful change cannot occur. These barriers, in more detail, are:

- Lack of urgency (management must see the need for change);
- Lack of support for new approach;
- No vision statement (employees do not understand the direction in which the company is moving);
- Inadequate reward system (a reward system based on department goals and not company direction);
- Lack of communication and knowledge sharing (departments capture and guard knowledge);
- Lack of ownership among employees (employees do not feel they contribute to the company);
- Lack of customer involvement (voice of customer ignored); and
- Lack of supplier involvement.

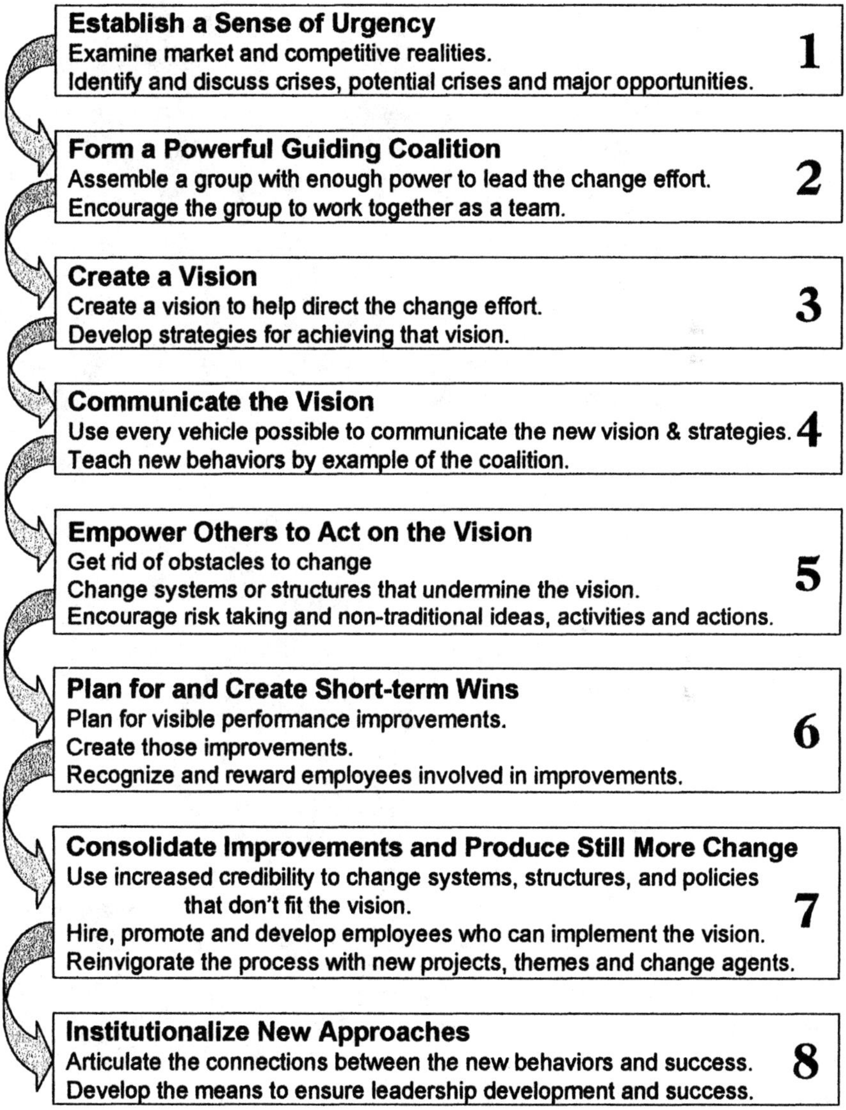

Establish a Sense of Urgency 1
Examine market and competitive realities.
Identify and discuss crises, potential crises and major opportunities.

Form a Powerful Guiding Coalition 2
Assemble a group with enough power to lead the change effort.
Encourage the group to work together as a team.

Create a Vision 3
Create a vision to help direct the change effort.
Develop strategies for achieving that vision.

Communicate the Vision 4
Use every vehicle possible to communicate the new vision & strategies.
Teach new behaviors by example of the coalition.

Empower Others to Act on the Vision 5
Get rid of obstacles to change
Change systems or structures that undermine the vision.
Encourage risk taking and non-traditional ideas, activities and actions.

Plan for and Create Short-term Wins 6
Plan for visible performance improvements.
Create those improvements.
Recognize and reward employees involved in improvements.

Consolidate Improvements and Produce Still More Change 7
Use increased credibility to change systems, structures, and policies that don't fit the vision.
Hire, promote and develop employees who can implement the vision.
Reinvigorate the process with new projects, themes and change agents.

Institutionalize New Approaches 8
Articulate the connections between the new behaviors and success.
Develop the means to ensure leadership development and success.

Figure 2.3 Steps for a successful organizational change [Kotter, J., "Leading Change: Why Transformation Efforts Fail," *Harvard Business Review*, March-April 1995, pp. 59-63. Reprinted with permission from Harvard Business School Press].

2.3 STEPS TO SUCCESSFUL CHANGE

There are a number of steps that must be taken to ensure a successful implementation of an organizational change such as the adoption of CE. A map, developed by Kotter [1995] and shown in Figure 2.3, was developed after examining more than 100 companies. In this map, Kotter outlines the eight steps for transforming an organization. In using these steps, several issues must be addressed [Kotter, 1995]:

- Change must be broken into stages, which can take considerable time to complete;
- Eliminating steps may seem to increase the speed of change, but may actually slow the process due to a lack of understanding at key places along the process;
- Critical mistakes along the way can set back the whole project or stop it entirely, so care must be taken to follow the steps and communicate with all those involved.

2.4 SUMMARY

There are three components that a company must examine when considering a major change in management philosophy such as the adoption of CE: the organizational structure of the company, personnel practices, and business practices and procedures.

- The functional product development structure is most often used by small companies. However, the structure that a company chooses to use should be dictated mainly by the type of product to be designed: innovative or follow-on.
- The use of empowered, cross-functional teams is important to the success of the company and the reduction in time-to-market. Teams are the only way in which true integration across functions can occur.
- Team empowerment is one of the most difficult ideas to implement, because it means that managers will lose some measure of control they once had over the team. However, better decisions can be made about the design of the product since the decisions are opened to those with the most information. Managers will then be freed to concentrate on more global and strategic issues rather than the day-to-day decisions that can slow down the design.
- Most of the barriers to the implementation of CE can be addressed by leadership of the senior management. Senior management must set the direction, support the new approach, reward those implementing new procedures, and communicate the successes with all employees.
- Changing to a new management philosophy will take time and should be performed step-wise.

REFERENCES AND BIBLIOGRAPHY

ATK/Bay Area Incorporated Mission and Vision Statements, http://www.atk-engines.com/mission.html. ATK Engines, 3210 South Croddy Way, Santa Ana, CA. 92704.

Block, P., "The Empowered Manager," *Positive Political Skills at Work*, Jossey-Bass, 1987, p. 115.

Coger, J. and R. Kanungo, "The Empowerment Process: Integrating Theory and Practice," *Academy of Management Review*, Vol. 13, pp. 471-482, 1988.

Ionics, Incorporated http://www.ionics.com/aboution/mission.htm. Ionics 65 Grove Street, Watertown MA, 02472-2882.

Katzenbach, J. and D. Smith, *The Wisdom of Teams: Creating the High Performance Organization*, Harvard Business School Press, Boston, 1993.

Kotter, J., "Leading Change: Why Transformation Efforts Fail," *Harvard Business Review*, March-April 1995, pp. 59-63.

Stetter, R. and D. Ullman, "Team Roles in Mechanical Design," *Proceedings of the 1996 ASME Design Engineering Technical Conference and Computer in Engineering Conference*, 96-DETC/DTM-1508, August 18-22, 1996, Irvine, CA.

Syan, C. and U. Menon, *Concurrent Engineering: Concepts, Implementation and Practice*, Chapman & Hall, 1994.

Wheelwright, S. and K. Clark, *Leading Product Development: The Senior Manager's Guide to Creating and Shaping the Enterprise*, Free Press, New York, 1995.

3

A Concurrent Engineering Methodology

In collaboration with Hans-Peter Kemser, Johanne Phair, and Chris Henson

In this chapter, we will overview the concurrent engineering methodology, which includes the product development model and the production process development models. These models use concurrent engineering as their overriding philosophy, strengthened by the use of a systematic design process. We will also give an overview of all the phases, steps, and team members of product and process design. In later chapters we will discuss the individual models in more detail.

3.1 AN OVERVIEW OF CE METHODOLOGY

CE methodology can generally be divided into five phases as shown in Figure 3.1. The methodology begins with an initial product idea during the product planning phase. Product design development and production process development occur simultaneously. The type of development project and the existing manufacturing processes determine the start of these phases, which may occur together or more likely will occur at different times. For new production processes, the start of production process development will often precede product development. If existing production processes are in place, and a new product is to be developed, the start of product development will precede process development.

Figure 3.2 shows a schematic of the CE methodology in more detail and includes phases and milestones. As shown in this figure, a kick-off meeting is held, signifying the start of the project. All product and process team members along with management, are present at this meeting in which the project goals are

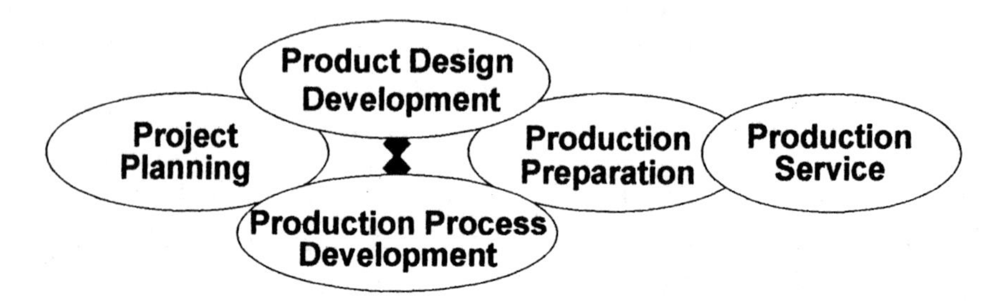

Figure 3.1 An overview of CE methodology phases [Reprinted with permission from H. Kemser].

defined. The figure also shows that product and process design may be subdivided and undertaken in parallel. As stated above, these design processes do not necessarily begin at the same time, but rather the timing depends on the amount of development needed. However, these development efforts should be timed to finish together.

It is important to clarify the terminology used throughout this book, and the terminology is best explained through Figure 3.2. The integrated design process shown in the figure is known as the concurrent engineering methodology. On the left side of the figure is the product development model, and on the right side are the process, testing, and packaging development models. The figure also shows multiple design and process development models being undertaken in parallel. The shaded regions within the models are known as phases, and the phases are further divided into steps. Each phases ends with a milestone shown in the figure as a diamond. A milestone is the point at which all steps and outputs from the preceding phase should be completed and management approval is sought. The following sections will explain this methodology in more detail.

3.1.1 The Product Development Model

The product development model is made up of four phases as can be seen in Figure 3.2: project planning, conceptual design, design, and production preparation. Figure 3.3 shows the product design model, and all of its phases and steps. Between each of the four development phases, a milestone indicates when management should approve the progress of the project, ensuring that all work and outputs are completed at the end of the phase. The four milestones in the design model are defined as project approval, program approval, design approval, and customer and production approval. The four phases are divided into design steps. For example, the project planning phase is divided into identify needs, define product specifications, and plan development tasks Each of these phases and steps with be discussed in detail in the next chapter, along with their associated outputs.

3.1.2 The Production Process Development Model

The manufacturing development model actually consists of three different models that can be used in any combination. The process development model defines the steps necessary for the design of the manufacturing line that will be used to produce the product. The testing model specifies the development steps for the design of the necessary testing tools and software needed to ensure that the manufactured product meets the quality and performance standards established by the company. Finally, the packaging model defines the steps for designing the process or line used to package and prepare the product for shipment. These models work concurrently with one another as well as with the product development model. Concurrent development allows the design of the individual processes and equipment to occur simultaneously, while communicating progress between the teams to ensure that the product can be manufactured, tested and packaged easily.

The process model is shown in Figure 3.4; the testing and packaging models which are very similar are shown and discussed in Chapter 5. This model, as also seen in the product development model, has four development phases. The first phase is the planning phase that is common among all the models, and is used to coordinate the project goals, the schedule, and the resources between and among the various teams. There are four milestones for each of the three models: project approval, concepts approval, design approval, and customer/production approval. Each of the four phases is divided into development steps as seen in the figures. These steps and their outputs will be discussed in more detail in Chapter 5. All the models, including product development, converge at the onset of the production/service phase.

As discussed in the previous chapter, cross-functional teams are the backbone of concurrent engineering methodology. The following section introduces the team structure, which can differ from a company's general operational structure. Because small companies have fewer resources than large corporations, employees are usually required to work on multiple teams, meaning that they will be working on multiple projects, and that they will be required to have multiple perspectives of the development process throughout the project.

3.2 PRODUCT AND PROCESS DEVELOPMENT TEAMS

One of the primary attributes of concurrent engineering is the use of cross-functional teams which provide a life-cycle focus to the project.. There are five teams that participate in the concurrent engineering methodology: the project management team, the new technology team, the cross-functional product

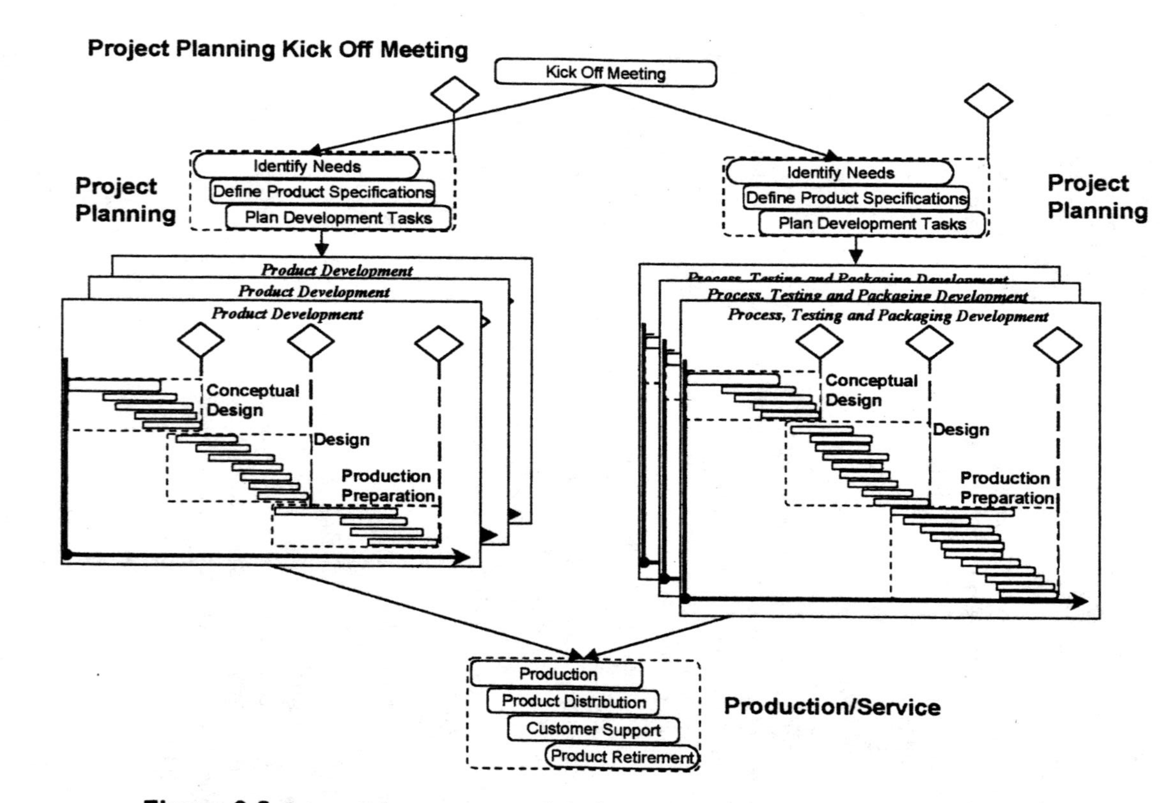

Figure 3.2 Integrated concurrent engineering methodology [Adapted with permission from J. Phair].

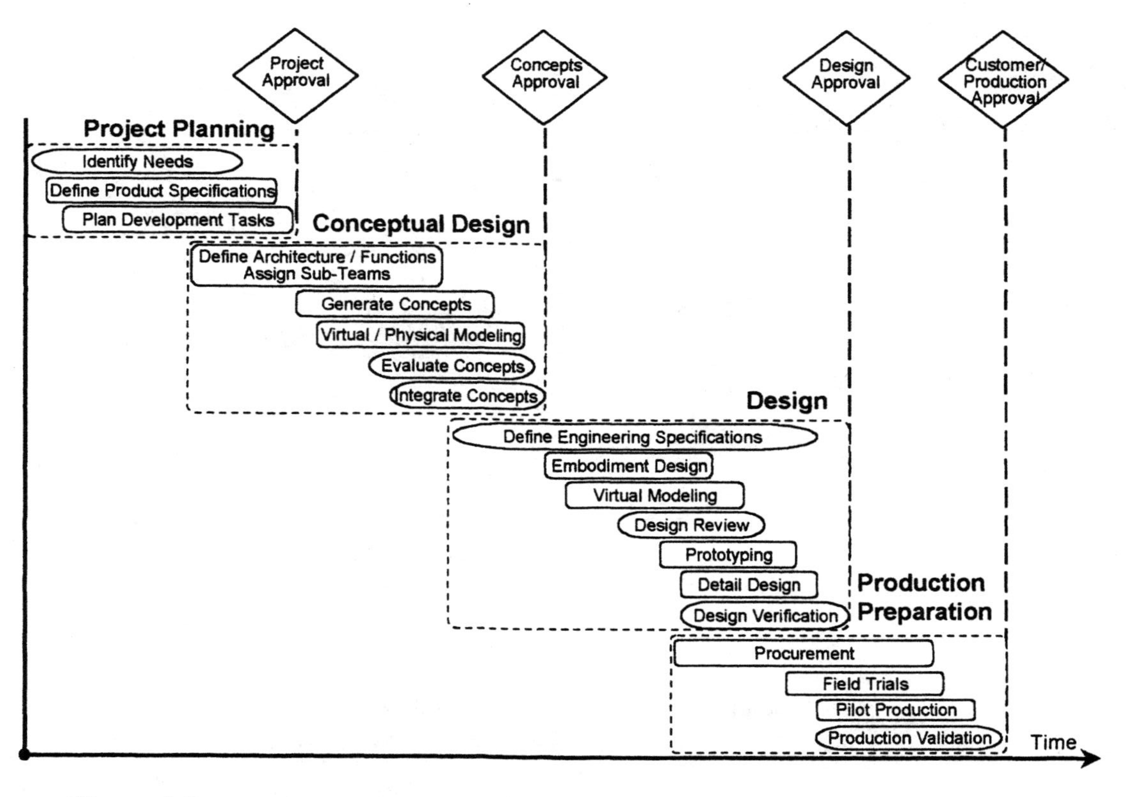

Figure 3.3 The product development model showing phases, steps, and milestones [Reprinted with permission from H. Kemser].

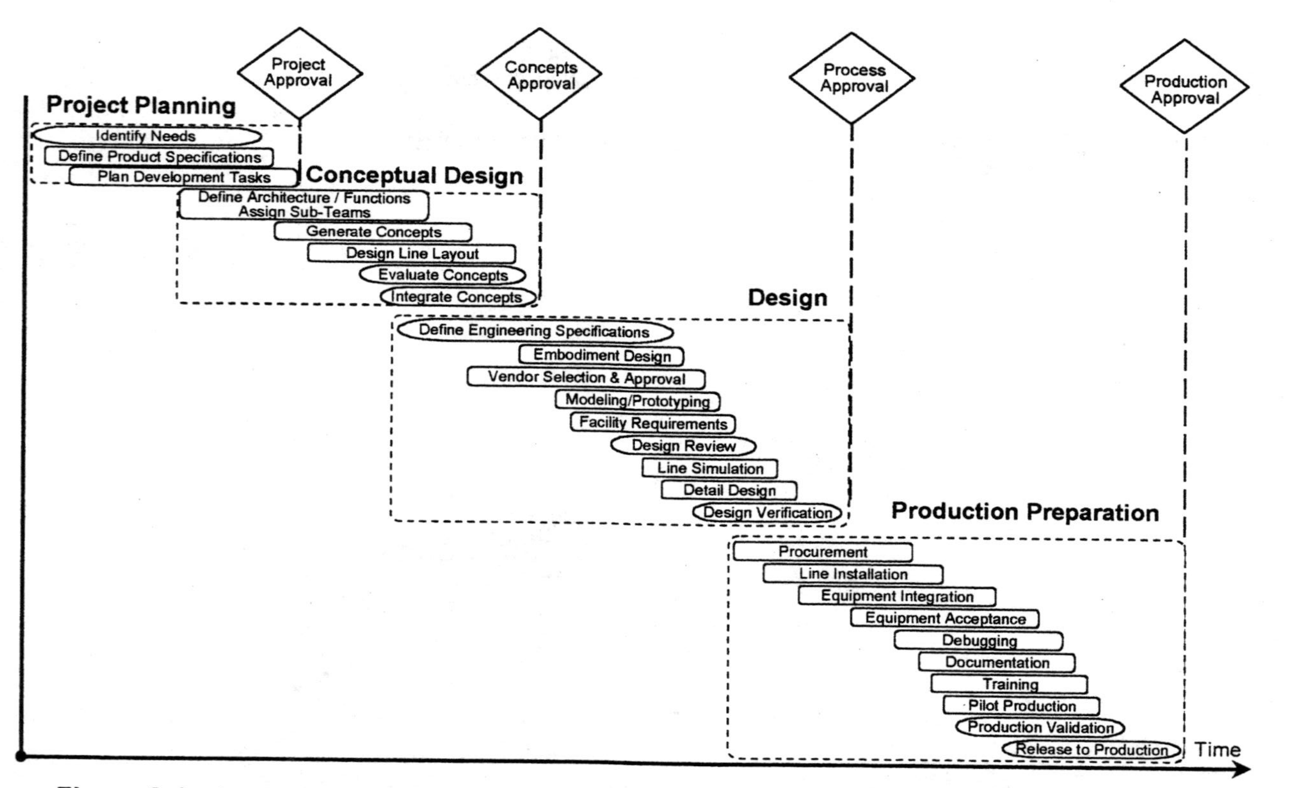

Figure 3.4 The manufacturing process development model showing phases, steps, and milestones [Adapted with permission from J. Phair].

development team, the cross-functional process development team, and the product team. Since a company's size and product lines mainly define the responsibilities and members of each team, the following is a general overview of the team activities, responsibilities, and recommended members throughout the project. These teams will vary in size depending on the size of the company and the requirements of a particular project. It is possible to be a member of more than one team, particularly in smaller companies where a designer may be designing a portion of the manufacturing line as well a part of the test and packaging lines. Also, a designer may also be working on multiple projects simultaneously, as well as being a member of different teams on the same project such as being a member of the new technology team as well as the product development team. Therefore, the responsibilities of the teams can vary depending on the project and should be adjusted accordingly. Figures 3.5-3.8 illustrate what phases and tasks the teams are responsible for, and how they work with one another as seen through their overlapping responsibility bubbles.

3.2.1 The Development Management Team

The development management team (MT) generally consists of senior management personnel, but can vary based on the size and structure of the company. The MT is active in the project planning phase and in all the milestones throughout the project, both in product and production process design development. The MT sets priorities, appoints team leaders, allocates resources, approves the timeline of the new projects, forms cooperative partnerships with suppliers or other companies for the development of portions of the designs, approves budgets, and approves of single-source suppliers. Although this team has oversight for the project, most decisions and all recommendations are made by the various development teams.

3.2.2 The New Technology Team

The new technology team (NTT) is made up of various design and manufacturing engineers from throughout the company. This team is responsible for staying informed about new technologies and advances, especially those related to their industry. It is not the mission of the NTT to develop new technologies within the product development process, as this leads to uncertainty in the performance of the product and its delivery. Instead, the invention of a new technology or process should be separated from the development of a product, thereby avoiding the development process being driven by the technology rather than by the market [Clausing, 1994; Wheelwright and Clark, 1992]. Rather, this team develops new technologies outside of the development process, and brings those that are proven to the development process. The main tasks of this team are to ensure that a) the latest technologies can be feasibly provided at the initiation of the project, b) many different technologies are explored without causing scheduling delays, c) new technologies are being developed and will be available if the company must react

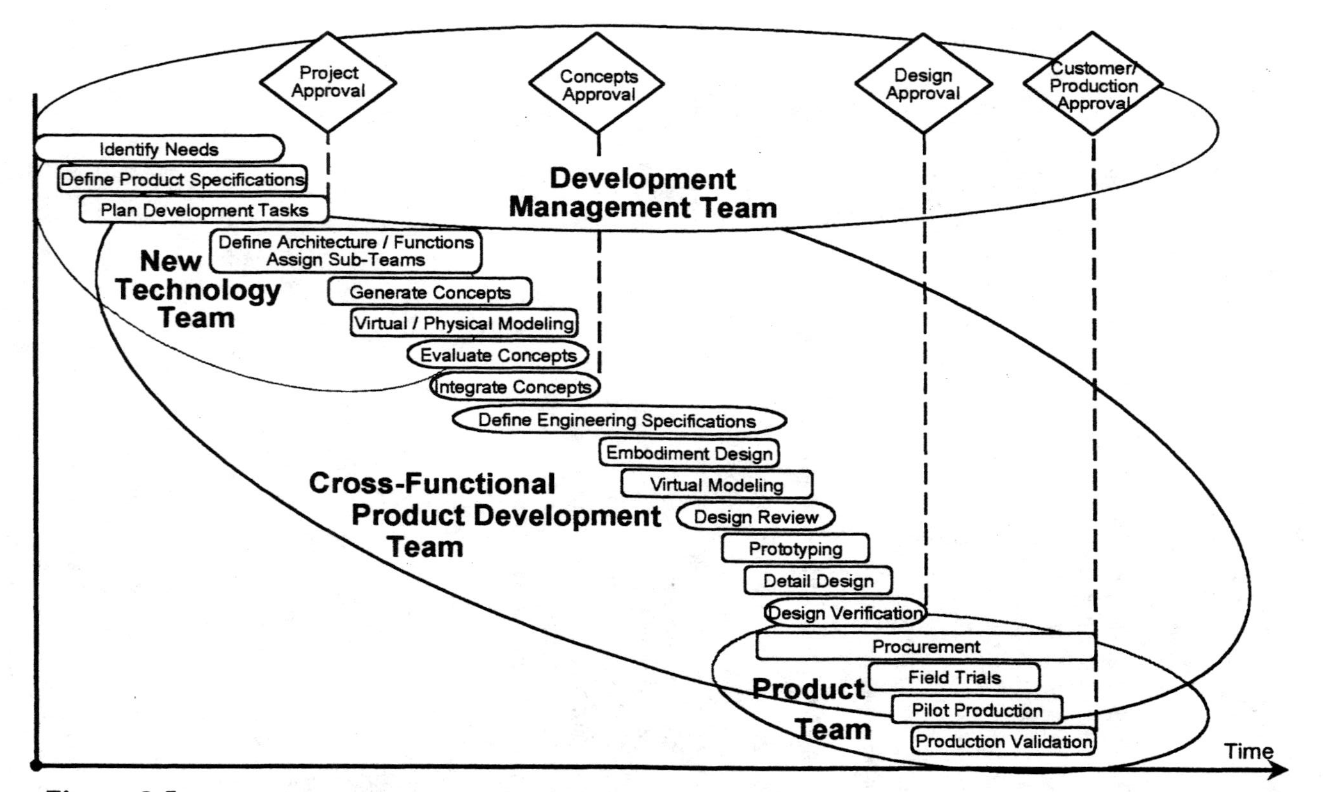

Figure 3.5 The product development model showing team responsibilities [Reprinted with permission from H. Kemser].

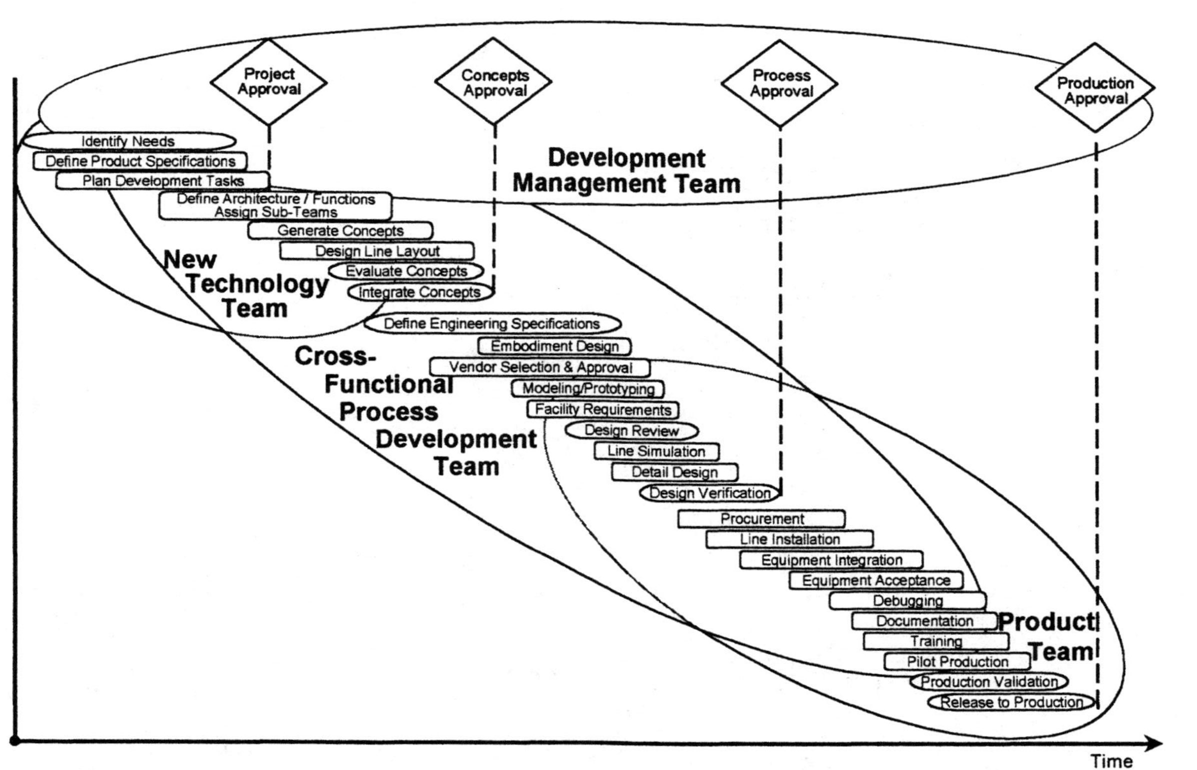

Figure 3.6 The manufacturing process development model showing team responsibilities. [Adapted with permission from J. Phair].

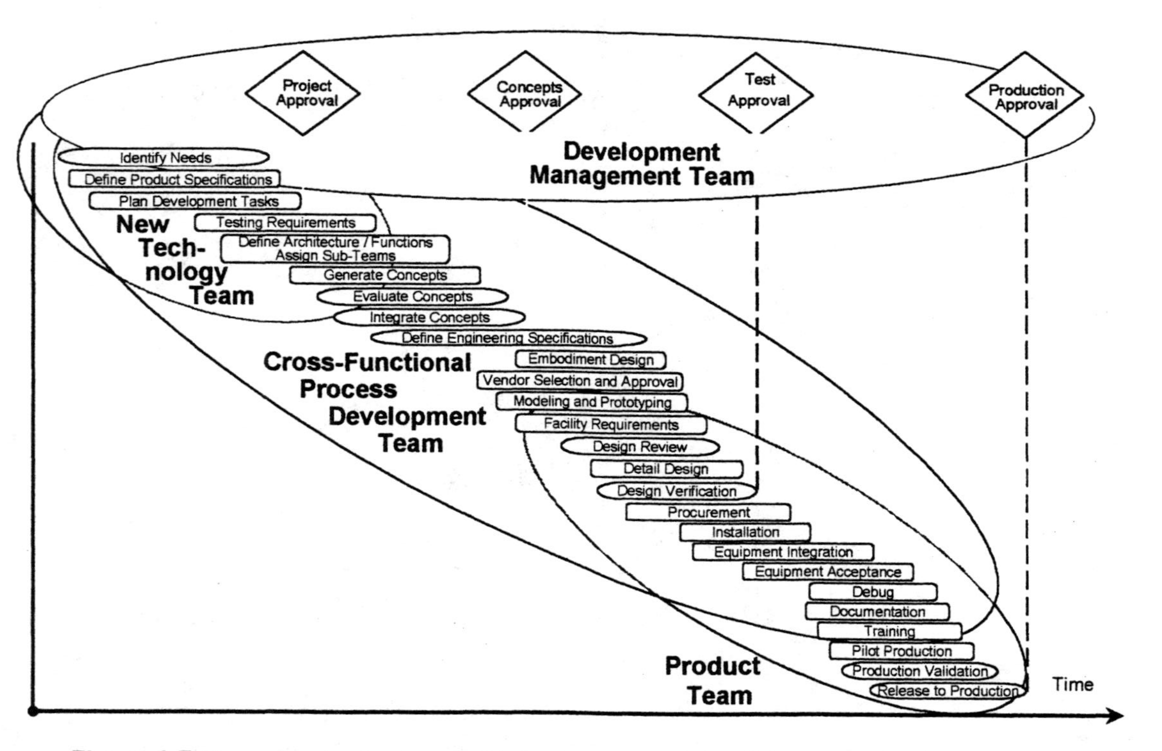

Figure 3.7 The testing development model showing team responsibilities [Adapted with permission from J. Phair].

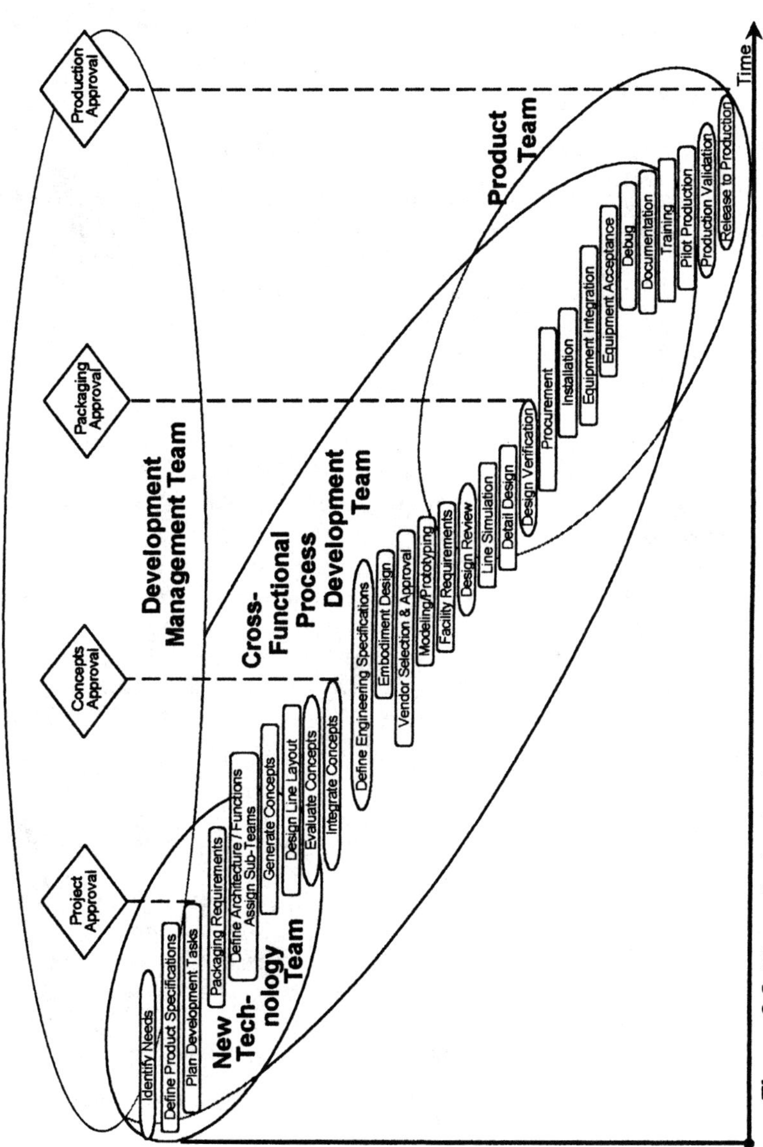

Figure 3.8 The packaging development model showing team responsibilities [Adapted with permission from J. Phair].

quickly to market changes. This team is active during the project planning phase and assists the development teams during the conceptual design phases. Since small companies are generally limited by resources including money as well as personnel, the NTT usually consists of members who are also on the product or process development teams.

3.2.3 Cross-Functional Product Development Team

The cross-functional product development team (CDT) includes members from different functions within the company, thereby representing all phases of the product's life cycle. For example, the CDT could include representatives from sales and marketing, manufacturing, test, field service, procurement, and finance, in addition to design and manufacturing engineers from different functional areas that could include software, electrical design, and mechanical design. Furthermore, if the project includes a partnering company, the team will have members from both companies. In many small companies, the members of the CDT might fill more than one position within the team depending on their responsibilities.

The team begins its activities during the planning phase, and continues through all phases until it transfers the project to the product team during the production preparation phase. Concurrent engineering implies that activities occur concurrently. The CDT may break into sub-teams to perform the conceptual design and design phases simultaneously depending on the architecture of the product. The use of parallel teams was illustrated in Figure 3.2. Again, in small companies, a team may be made up of just one or two people, and the team members are usually on multiple teams and must divide their time between the tasks. Communication among the CDT teams is critical to ensure that the design can be integrated. In addition, it is important that communication occur frequently with the cross-functional process development team to reduce problems that occur in the production preparation phase.

Using this concurrent approach can significantly reduce the overall development time. However, this type of development can be successful only if a) communication occurs within the teams, b) communication occurs between the teams frequently, and c) the project manager ensures that the components are integrated as early as possible in order to work out any problems. Integration of the design will be one of the most challenging aspects of the project. Ideally, the transfer of the product to production will occur with few problems because of this early communication between teams and because manufacturing engineers are members of the CDT.

3.2.4 The Cross-Functional Process Development Team

The cross-functional process development team (CPDT) is responsible for the development and installation of the manufacturing line. Just as in the CDT, the CPDT can be divided into sub-teams that will work on different aspects of the line simultaneously, such as manufacturing line design, test design, procurement, packaging design and line design, and research and development of new technologies. Again, for success, communication among and between teams is imperative. Without communication, tracking changes is difficult and can result in scheduling delays. The CPDT works in parallel with the CDT throughout the concurrent engineering process. The timing of the start of these teams varies with the amount of development work required. This issue will be illustrated in Chapter 6.

3.2.5 The Product Team

The product team (PT) is formed from members of both the CDT and the CPDT, as well as from members of other PTs. This team focuses on the support of the product in the production/service phase, addressing any problems that arise either with the product in the field or with the production lines. Ideally, because field service and manufacturing is represented in the CDT, few problems with the product will arise, which is another advantage of the concurrent engineering process over a sequential design process. Any problems that arise with either the product or the production lines should be documented and relayed to other projects in the design phases, as these problems may be avoided in new designs and follow-on products. The PT membership should rotate among the CDT and CPDT members. Ideally, every third project a design engineer should serve a term on a product team. This rotation accomplishes several things:

- Designers that are transferred to PTs retain ownership of the design, thereby increasing the desire to design it right in the first place.
- Knowledge gained in manufacturing the product and supporting the product are transferred back to the design teams.
- "Firefighting" problems that arise with the design and manufacturing lines are removed from design engineers remaining with the PT. This frees the design engineers to concentrate on the new design project rather than solving problems with "old" designs.
- Finally, the rotation eliminates the idea that product support is a second class job and lower in importance than the design of new products.

In our observation of many small companies, in those without a PT, the engineers spent a significant portion of their time supporting products in the field (> 30%). However, in those companies with a PT, the engineers spent less than 10% of their time with product support. This team plays a key role in improving morale as well as future products.

The team structure presented above will, in general, differ from the company's general business operating structure, as was discussed in Chapter 3. Both the development team structure and the company's general operating structure are in place simultaneously. For example, if the company follows a functional or a lightweight functional organization, the product development teams would be formed from members of the functional areas. These team members would be expected to perform as development team members as well as to continue fulfilling their functional activities.

3.3 DEVELOPMENT PHASES

As discussed in Section 3.1, there are four development phases for both the product and process development models, and then they converge at the fifth phase of production/service. Each phase has the same structure as shown in Figure 3.9. The team members who are active in the phase, and the support members who provide additional input, are shown on the left. The design steps within the phase are shown in the middle, and any step shown as an oval is defined as a communication port. A communication port indicates that it is crucial that the product and process design teams meet to discuss the decisions made within this step, as the decisions made will affect all teams. Ideally, the teams should communicate frequently, if not daily, but a communication port ensures that at the very least, communication occurs at these critical decision points. On the right of the figure are the outputs, which are the decisions and designs that result from the design step. As indicated, the outputs should be recorded in a database for team members to access during the concurrent engineering process as well as in the future on later projects to understand the history of a product line. Between the communication environment and the design steps, there is an iteration pipeline. This pipeline shows there will likely be iteration within a design phase before that phase is completed, as decisions are made and revisited as the teams meet to discuss and integrate their designs. Figure 3.10 provides a key to the team and support members found in the communication environment in Figure 3.9.

The process of moving through and completing each design phase can be imagined as a three-dimensional spiral, in which information within the spiral progresses from the general to the specific. The center of the spiral represents the milestone of each phase. The work in each phase begins in a general fashion. With each step and with each iteration, the team increases the specificity of the decisions and the design, narrowing the initially wide boundaries of the design. For example, the planning phase begins with the definition of the identification of needs. In this step, the teams discuss general ideas and requirements for the product or the manufacturing line. For example, if the team were designing a printer, they would indicate the number of pages per minute, whether it would be black-and-white or color, etc. The manufacturing team would want to know how many printers were to be made per day. After these general requirements were put forth, the teams would define the product specifications and begin to plan the development tasks. With each step, the teams move toward the center of the spiral, with the goal being the approval of the phase by the management team by way of a milestone. The next two chapters will provide a description of the steps in the product design and the production process design models.

Team(s)
Support
Product Development Phase
Output
Communication Port
Iterations
Development Step
Communication Environment
Development Environment
Database

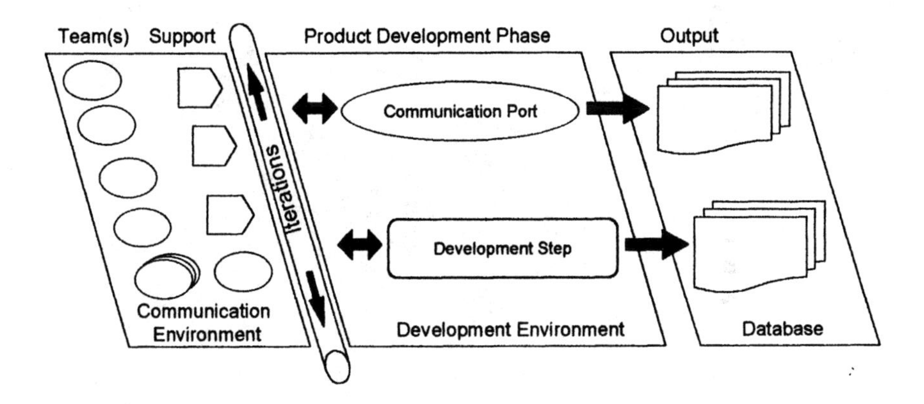

Figure 3.9 General layout of the development phases [Reprinted with permission from H. Kemser].

Team(s)
Support
MT
Development Management Team
NTT
New Technology Team
CDT
Cross-Functional Product Development Team
CDT
Parallel Cross-Functional Product Development Team
CPDT
Cross-Functional Process Development Team
CPDT
Parallel Cross-Functional Process Development Team
PT
Product Team
C
Customer
S
Supplier
FS
Field Service

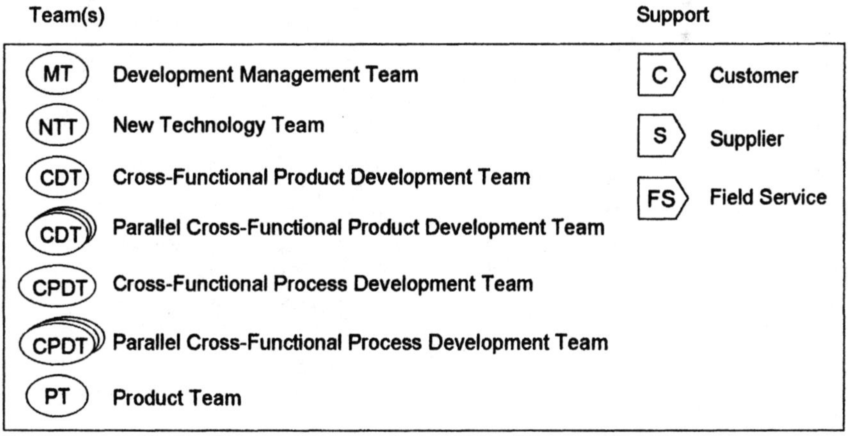

Figure 3.10 Key to team and support members [Reprinted with permission from H. Kemser].

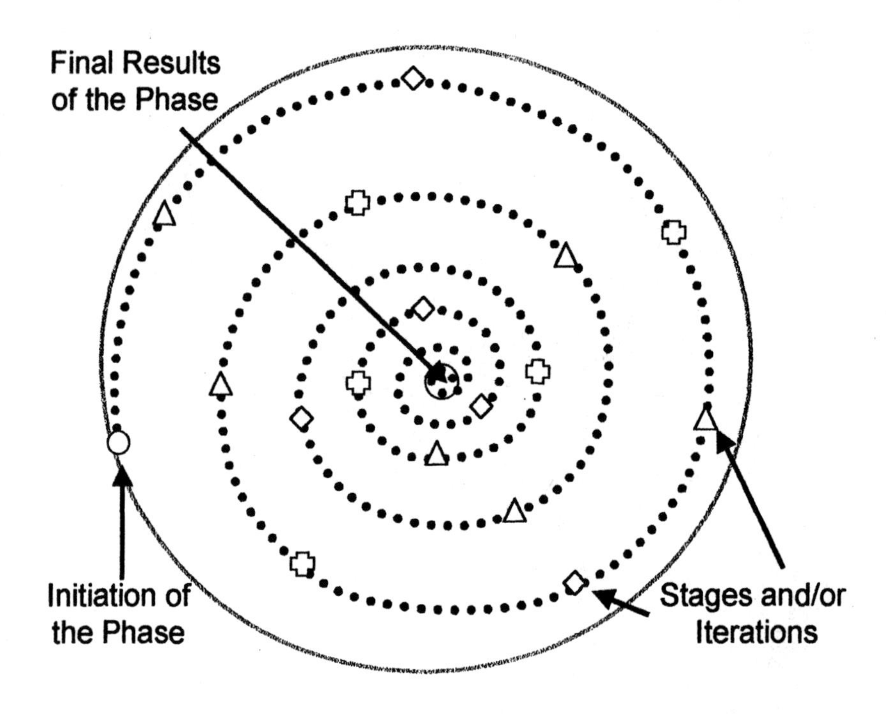

Figure 3.13 Development spiral for each development phase [Reprinted with permission from H. Kemser].

As discussed above, at the end of a design step, the team members document the decisions and ideas generated in the output. To ensure that the information is easily retrievable, a consistent structure must be maintained throughout the concurrent engineering process. The structure includes four sections: the title, the definers, the overview, and the documentation. With this structure, the user can decide how much information to view at one time, stopping at the overview, or digging into the documentation. The title gives the user an idea of what is contained within the output, and is therefore an important first look at the output. The overview contains all the information critical to understanding what has been accomplished in the associated design step. For example, the output for the field trials design step in the production preparation phase is the list of changes. In this output's overview, all of the changes made to the design should be listed in chronological order. The documentation section would then detail the rationale behind these changes and the alternatives to these changes discussed or tried by the team. In addition to the rationale, this section would also include the reasons, benefits, and costs of the change as well as any lessons learned.

There are four persons or groups involved in an output. The first person or group decides what type of information is needed for each design step output. This group will be called *creators*. The second person or group are called the *definers*. This group is responsible for developing the information required in the output, decisions, ideas, and designs. The third group is the *support team*, which gives any needed input to the definers. The fourth group is the *users*. There are four types of users: the definers, other team members, the management team, and future designers [Ullman, 1994]. The definers use the information in previous steps to develop the current output. Other team members will use the outputs to keep up to date with the project and ensure the decisions made will not impact their portion of the project negatively. The management team uses the output to monitor the project's progress. Future design team members can use the information to build upon past successes and avoid past mistakes and errors.

The most convenient way to store the outputs is a database, which can provide the easy accessibility. Data entry should be quick and easy to ensure that the team members use the system to record their major decisions made each day and the rationale behind those decisions. The data should be made available to appropriate team members during the design process, although read/write authority should be examined carefully. The outputs should be read/write accessible throughout the design phase until the approval of the milestone, at which time it should become read accessible only. The database should also track the completion of sections of the output so that the project managers and the management team can track the progress of the project. The database should also have search capability to allow team members to find any information that they need quickly. Finally, with a database, the computer can allow different levels of access to different types of people such as the customer, supplier, team members, and managers.

3.4 SUMMARY

Concurrent engineering methodology primarily consists of the product development model and the production process models, which include manufacturing line design, packaging line design, and testing design. These development processes are carried out by multiple teams: the management team, the new technology team, the cross-functional development team, the cross-functional process development team, and the product team. Because small companies have fewer resources, their employees are expected to serve on multiple teams and may be required to have different perspectives of the design, such as development and maintenance, while serving on the same team.

Each development model is divided into design phases, which are further divided into design steps. Each design phase follows the same general layout in which teams and support members are assigned to be responsible for various design steps. These teams then carry out the design step, which results in decisions and outputs for that step, which are recorded in a database. The decisions made within each phase begin very broad, but with time and iteration within the phase, the decisions become more detailed until a final milestone report for that phase is completed and ready for management approval.

The key benefits to this structured approach are that decisions are made in a timely, logical fashion and all team members and management are informed of these decisions. The entire life-cycle of the product is considered up-front in the design process, thereby eliminating problems such as designs that cannot be manufactured with the existing equipment. Also, the team members feel an ownership of the product and have a vested interest in seeing it successfully through manufacturing and production.

REFERENCES AND BIBLIOGRAPHY

Clausing, D., *Total Quality Development*, ASME Press, New York, 1994.

Kemser, H., *Concurrent Engineering Applied to Product Development in Small Companies*, Masters Thesis, University of Virginia, 1997.

Phair, J., *Integrating Manufacturing and Process Design into a Concurrent Engineering Model*, Masters Thesis, University of Virginia, 1999.

Ullman, *The Mechanical Design Process*, McGraw-Hill, New York,1994.

Wheelwright, S. and K. Clark, *Revolutionizing Product Development: Quantum Leaps in Speed, Efficiency and Quality*, Free Press, New York, 1992.

4

Product Design: Steps and Tools

In collaboration with Hans-Peter Kemser, Kevin Allen, and Chris Henson

Now that you have seen an overview of the product development model and the process development model, we will now discuss both in depth in this and the next chapter. Both models use concurrent engineering as their overriding philosophy, strengthened by using a systematic design process. First, we will discuss all the phases and their steps for product and process design. In Chapter 6 we will discuss how to tailor the models based on your company's particular needs and environment.

As discussed in Chapter 3 and shown in Figure 4.1, the product development model has five development phases: project planning, conceptual design, design, production preparation, and production/service. Between each of these five development phases, there is a management milestone, which is an approval point for the management team to assess the progress of the project. The four milestones are defined as project approval, program approval, design approval, and customer and production approval. Each of the phases are broken into steps. In this chapter, we will discuss each phase of the models and each step within the phases.

The first step in the design process is the kick-off meeting, an initial meeting of senior management and team leaders for the cross-functional product and the cross-functional process teams. These leaders will choose the team members for the project. In this initial meeting, management and the team leaders define the broad outlines of the impending project. The teams will then fill in these outlines in the subsequent development phases.

Table 4.1 Output for the design step: identify needs.

Title	Mission statement	
Definers: Primary	New technology team, cross-functional product development team, management team	
Definers: Support	Customer, field service	
Overview	Mission statement	
Documentation		
Company	General issues	Available resources
		Type of investment required
		Partners to provide skills or equipment not possessed by company
		In-house or purchased components
		Regulations or standards to be met
		Review past projects to look for areas of improvement, problems to avoid, successes to repeat
	Company strategy	Type and amount of risk company is willing to undertake
		Type of risk associated with project
		Fit of project with existing product lines
		Fit with operating strategy
		Fit with vision, mission, and goals of company
	Marketing	Market surveys: do we have them or do they need to be conducted?
		Expected time-to-market
		Window of opportunity to announcement date of this product
		Risks to company if announcement date is missed
		Targeted price range
Product needs	Customer	Customer requirements and expectations
		Product benefits
		Product usability: more, less, the same
	Competition	Competitive products currently in the market
		Key features of competitive products
		Price range of competitive products
		New technologies recently introduced
		Anticipated new technologies
		Product differentiation
	Marketing	Forecast for sales volume
		Customer complaints about our product line

contained in the documentation section will help the team do two things: 1) define the customer needs for a House of Quality which will be included in the milestone report, and 2) define the mission statement for this project. Mission and vision statement were discussed in detail in Chapter 2. However, in the overview section, the team basically needs to state why they are undertaking this project and what they plan on accomplishing in the project.

4.1.2 Product Specifications

Product specifications broadly outline what the product will do and what features it will have. Ideally, these specifications should have target values associated with them. For example, if developing a new printer, state how many pages per minute the printer should print in draft mode and in letter quality mode. Keep these specifications as general as possible to avoid defining solutions at this early stage in the design process. Table 4.2 shows the output for this design step and the checklist in the overview section should help define the product specifications.

Table 4.2 Output for the define design step: product specifications [Adapted with permission from from Pahl, G. and W. Beitz, *Engineering Design*, Springer-Verlag, New York, 1996.]

Title	Product Specifications
Definers: Primary	New technology team, cross-functional product development team
Definers: Support	Customer
Overview: Specifications	Product geometry
	Product footprint
	Required product functions
	Optional product functions
	Required product features
	Optional product features
	Product movement
	Forces that the product produces
	Forces that the product undergoes
	Movement that the product performs
	Movement that the product undergoes
	Energy required
	Energy produced
	Transportation conditions
	Types of environments product will be used in
	Maintenance required
	End-of-life: recycling, remanufacture, disposal
	Target cost
Documentation	Rationale for requirements and associated values

4.1.3 Plan Development Tasks

The final step in this design phase is to plan and schedule the development tasks. In this step, a rough budget for the product and the teams should be defined. Also, a preliminary schedule should be defined, which would later be defined in more detail by the various team members. In general, this schedule can be based on the required announcement date of the product, and the various milestones can be defined back from that date. As an alternative, the schedule can be based on previous projects. However, this technique should be used with caution. If this development project is to be undertaken using concurrent engineering whereas previous projects used sequential engineering or some variation thereof, then the schedule will not match up as we saw in Figure 1.2 in Chapter 1. The planning and conceptual design phases will take much longer than they did in the past when using concurrent engineering, but this time should be made up in the detail design and production preparation phases since integration should be easier and changes to the design should be minimal in these phases.

Table 4.3 Output for the design step: plan development tasks.

Title	Project Plan
Definers: Primary	New technology team, cross-functional product development team
Definers: Support	
Overview:	
Estimate resources	Cost of development
	Required personnel: team members and support
	Shop facilities
	Computer equipment
	Equipment needed for development and testing
	Resources needed from partners or suppliers
Plan schedule	Estimate of development and production schedule
Documentation	Rationale for resource needs and schedule

Iteration among the steps within this phase is to be expected. As new information becomes available, then revisions to the needs, product specifications, or schedule may need to be made. The planning phase milestone is the most important of the design process. The decision to continue with the development process or abandon the project should be made at this point. If this go/no go decision is made at a later milestone, the costs are much higher for cancellation of the project as little investment has been made at this point in the project. Furthermore, this marks the beginning of a development project, not a research and development project. No untried technologies should be undertaken simultaneously with the development of a new product. Generally, the development and perfection of new technologies should take place outside of the development process, and then applied when the engineers are comfortable with them. However, if the design team feels comfortable with the new technology, they may decide to undertake its inclusion in the design. But, they should have a good sense of the risk the new technology poses, and have a back-up plan in case of failure.

All the key decisions made within this phase should be recorded as part of the outputs within a database. This database can be used by all team members during the project to check on progress, to get data and decisions relative to their piece of the development project, and to communicate with each other. Once a project is completed, this data can be used as a resource for other projects to help avoid mistakes made by past teams or to repeat their successes, and to help develop schedules and cost estimates.

This phase concludes with the completion of the project planning document which includes the mission statement, a product description, company issues including strategy and goals for this development project, the intended markets for the product, the competitive environment, the customer requirements, the product specifications developed from the needs, the budget, the required and requested resources, and the preliminary schedule. The most important question that the document must answer is why the company should undertake this development project, or in rare cases, why it shouldn't. The management team will review this document and, based on it, recommendations from the team leaders will decide whether to approve the document and project. With approval, the project can then move to the next phase.

An alternative form to listing the requirements and specifications in the project planning document is to use a graphical representation called the House of Quality (HOQ); an overview of this tool is shown in Figure 4.3. Unless the product is something that is completely original, and therefore customers will have no idea of its form or use, then an HOQ is a useful way to represent the voice of the customer and its relationship to the product specifications.

The HOQ was first used by the Japanese in the 1970's as a tool to improve Toyota's automobile rust record. Since that time it has been widely adopted in the U.S. as a tool to improve product quality and understanding of the product. It can be applied to the entire project or portions of it. In Figure 4.3, we have shown six "rooms" in the house, although the diagram can be even more involved [Day, 1993]. The HOQ is a time-consuming tool to use. However, it can be worth the time. It promotes better understanding of the project and of customer needs, and promotes better communication between team members. Finally, it can focus the team on the most important attributes of the proposed product that can make the product more competitive.

To develop a HOQ, we start with the customers' wants and needs. These wants can be determined through several different methods including customer

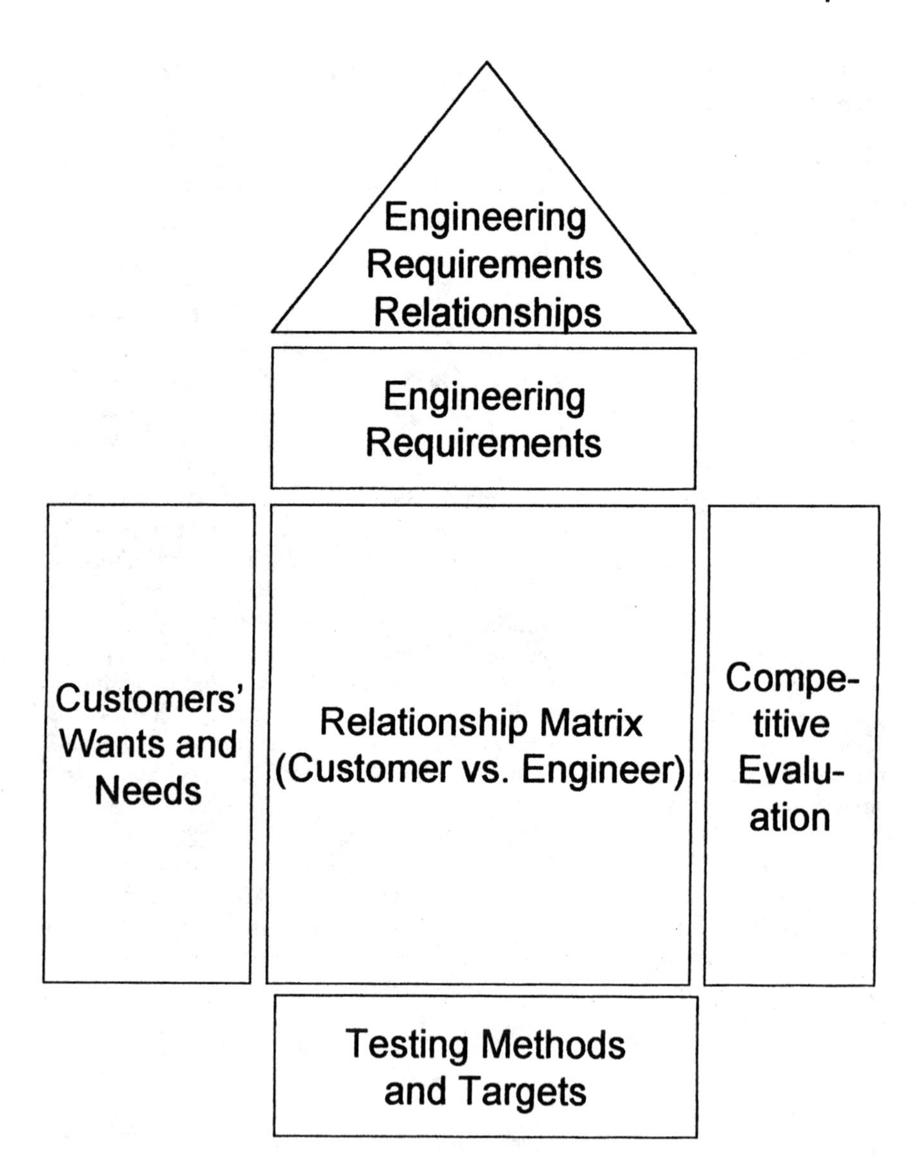

Figure 4.3 An overview of a House of Quality.

surveys, interviews focus groups, or observing product use. These wants should be rated as to how important they are to the customers using the following scheme:

5: Must have
4: Above average importance
3: Average importance
2: Below average importance
1: Not important

Next, we want to determine how the customers' wants can be described in measurable terms through the use of the engineering requirements. The customer attributes need to be refined into engineering attributes that can be felt or observed by the customers. These are the quantifiable specifications that were developed in the define product specifications design step.

The next section to be completed is the relationship matrix. This section shows the relationship between the customer attributes and the engineering attributes. To show this relationship, use the following symbols:

● Strong positive relationship (5 points)
O Medium relationship (3 points)
Δ Weak relationship (1 point)

Each customer attribute should have at least one relationship with an engineering attribute; to ensure all customer requirements are addressed. However, not all engineering attributes need to have associated customer attributes. Some engineering requirements may not be anticipated by the customer. For example, safety requirements are often neglected by the customers or are implicit in their product requirements. Therefore, every row should have at least one relationship symbol, but not every column must have a relationship symbol.

For the competitive evaluation, competitors products are assessed applying the customer attributes using survey data and comparing to the existing product or product line. The following scale is used:

5: World class
4: Very good
3: Good
2: Not very good
1: Poor

The next step is to set the target for the product to be developed for each customer attribute. These targets will help the team decide the attributes on which to concentrate to meet the desired product market position.

In the engineering relationship matrix, the engineering characteristics' relationships, e.g., how changing one engineering characteristic can affect others. The following scale is used:

✓: Strong positive relationship
+: Medium positive relationship
– : Medium negative relationship
X : Strong negative relationship

Finally, for the testing targets, a method is defined for measuring each engineering attribute and ideal targets are set for each of the measurements. See Appendix A for an example of a completed HOQ.

4.2 THE CONCEPTUAL DESIGN PHASE

The conceptual design phase consists of five design steps as shown in Figure 4.4. The phase begins with a definition of the product architecture and functions from which the assignment of sub-teams is made. Next, different design concepts are developed, modeling is performed, the concepts are evaluated and the best design concepts are selected for further development. Finally, the selected concepts are integrated into a set of final concepts. This phase ends with the approval of the final concepts by the management team through the concepts approval milestone.

4.2.1 Define Architecture/Functions and Assign Sub-Teams

This first task of the conceptual design phase is often performed simultaneously with the planning phase. As indicated by the name, concurrent engineering often means that the next phase is begun before the previous phase is completed. However, care should be taken not to delay the approval of the previous phase too long, since the further into the next phase the team works, the more expensive it will be

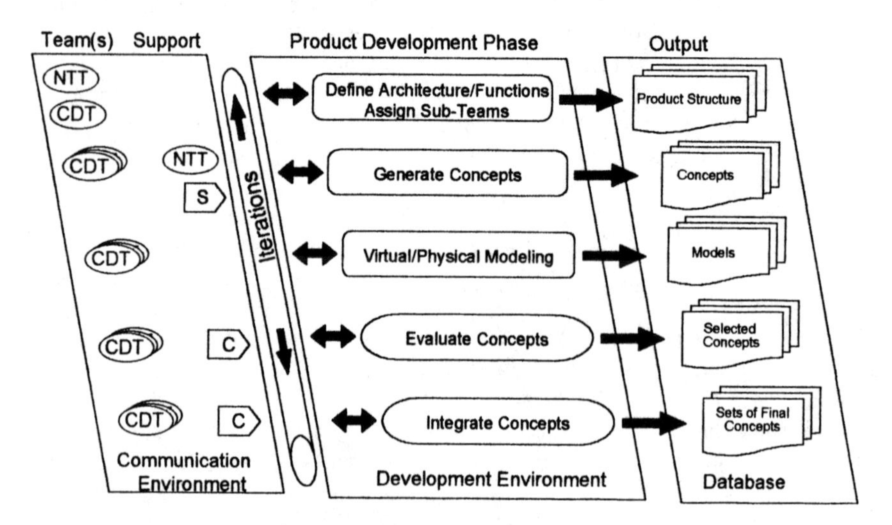

Figure 4.4 The conceptual design phase [Reprinted with permission from H. Kemser].

to cancel the project, if that is to happen. There are basically three types of designs that a company will undertake: original, evolutionary, or incremental. An original design is one that is either new to the marketplace or new to the company itself, that is, a design with no history. An evolutionary design is one that is based on a product already produced by the company, and an incremental design is one that entails minor changes to an existing design. These design types and their impact on the concurrent engineering process will be discussed in more detail in Chapter 6. However, at this point we can say that if the design type is original, then the new technology team (NTT) and the cross-functional product development team (CDT) will define the functions that this new design is to perform. On the other hand, if it is an evolutionary or incremental design, then the product architecture will have been previously defined. In either case, sub-teams will be assigned for the remaining steps and phases. The parallel development of the product can speed development, but can only be successful if there is communication within the sub-team, communication occurs frequently between the sub-teams, schedules are developed and maintained to ensure that tasks are accomplished on time, and that the sub-teams integrate their designs as soon as possible during the development process.

For an original design, the NTT and CDT must define the basic functions that the product must perform. The most common manner in which to represent the functions of a product is the function diagram in which the functions are linked by the flow of material, energy or information. First, the overall function of the product must be defined along with its associated inputs and outputs of materials, energy, and signals or information. Next, this overall function is broken into sub-functions. All functions are action words that define what the product does. How the product accomplishes the function is addressed in the next design step. The sub-functions are generally laid out in the order in which they will be accomplished. The sub-functions are linked by the flow of material, energy, or information between them. Figure 4.5 shows an example of a function diagram for an ice cream scoop. The large, dotted-line box shows the overall function of the scoop, and the smaller boxes within the overall function are the sub-functions. Material flows are shown in bold, broken lines. In this case, the material is ice cream. The energy flows are shown by the solid lines and represent both potential (PE) and kinetic energy (KE).

If the product is evolutionary or incremental in nature, then we must define its architecture based on the existing product. Product architecture is defined as the allocation of physical components to the various functions of the product [Ulrich, 1995]. In general, there are two types of architecture: integral and modular. An integral architecture is one in which one component often fulfills many functions and there are coupled interfaces between functions [Ulrich, 1995]. A modular architecture is the degree to which a product is composed of modules with minimal interactions between them. A module is defined as a component or group of components that can be removed from a product non-destructively, and pro-

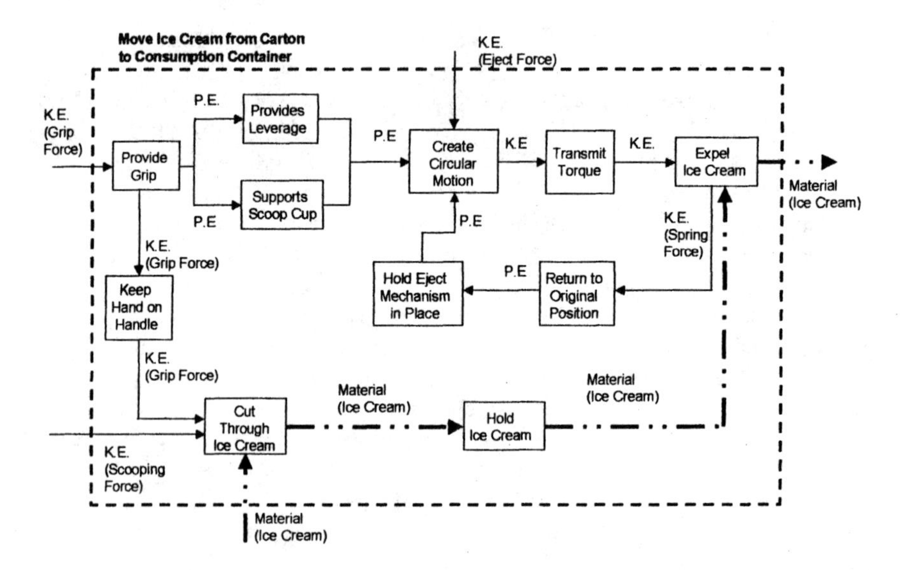

Figure 4.5 A functional diagram for a side-eject, ice cream scoop.

vides a unique basic function necessary for operation [Allen, 1998]. Product modularity is generally considered advantageous since modular products:

- Disassemble easily for reuse or recycling,
- Enable product variety easily,
- Allow component standardization,
- Reduce development time due to component use, and
- Correlate easily to team structures.

However, modularity does have its disadvantages, including:

- Increased production costs because of more assembly,
- Heavier products having greater volume,
- Possible reduction in quality due to more interactions and possible interferences, and
- Easier for competitors to copy.

The team must weigh these advantages and disadvantages to decide which type of product best suits their development project. Even so, a product cannot be either entirely modular or entirely integral, but will display more or less modularity. In the next couple of pages, we will introduce a technique for defining product architecture and assigning sub-teams. Figure 4.6 shows the steps for this technique.

The process shown in Figure 4.6 begins with the identification of the modules of the current product through disassembly. Next, the function structure for this product is defined, and from this information, sub-teams are identified. Then, the connections and flows between functions are defined. Finally, the modularity and the interactions metrics are calculated. Using this information, the CDT and the NTT can identify needed areas of improvement. In general, there will be iteration between these steps to ensure that the information is correct, and any changes made in later steps are correlated with earlier steps. Now we will discuss each of these steps in more detail.

First, identify the modules of the current product by disassembling it. A module is identified as a component or group of components that can be removed from the product as a unit non-destructively. This unit should provide one basic function that is necessary for the product to operate as intended. These modules, along with the identification of their basic functions, are defined as first pass modules and are documented on the right side of the reverse fishbone diagram [Ishii, 1996; Allen, 1998] as shown in Figure 4.7. If a module fulfills more than one function, then it has no basic function and must be refined. To refine the module, it must be disassembled further into sub-modules until the sub-module can be identified with a basic function. For example, if one of the modules removed from

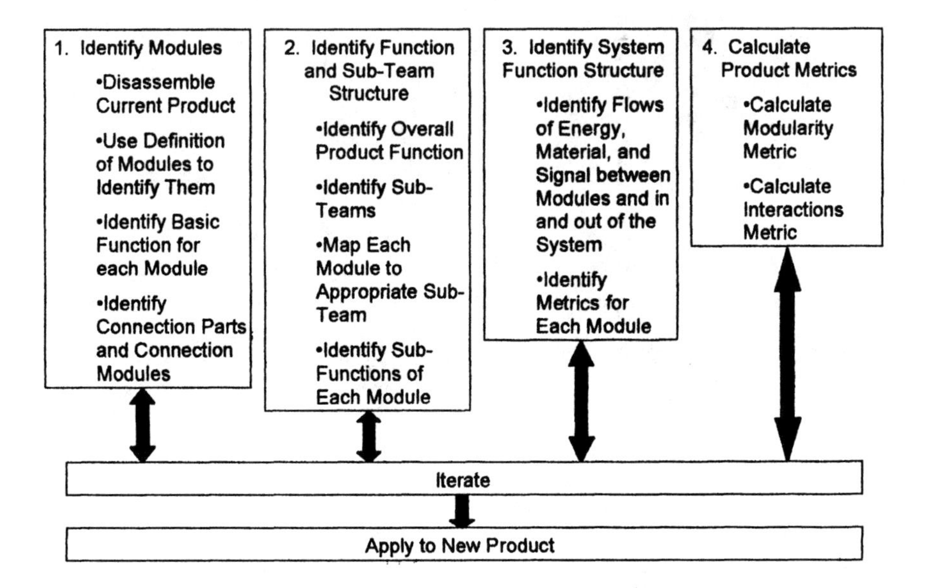

Figure 4.6 The steps for defining a product's architecture [Reprinted with permission from K. Allen].

a product is a panel consisting of a printed wiring board and a user interface panel, it must be further disassembled so that the panel and the printed wiring board are separated. Refined modules are connected to their associated first pass modules by a horizontal line and their functions are documented to the left of the spine as shown in Figure 4.7.

During disassembly, connection parts and modules must also be identified. Connection parts connect one module to another, and connection modules connect more than two modules together with multiple connectors. Connection parts are assigned to one of the modules that they connect. The connection modules are a special case that will be discussed more when we talk about the system function diagram. The connection parts are documented as shown in Table 4.4.

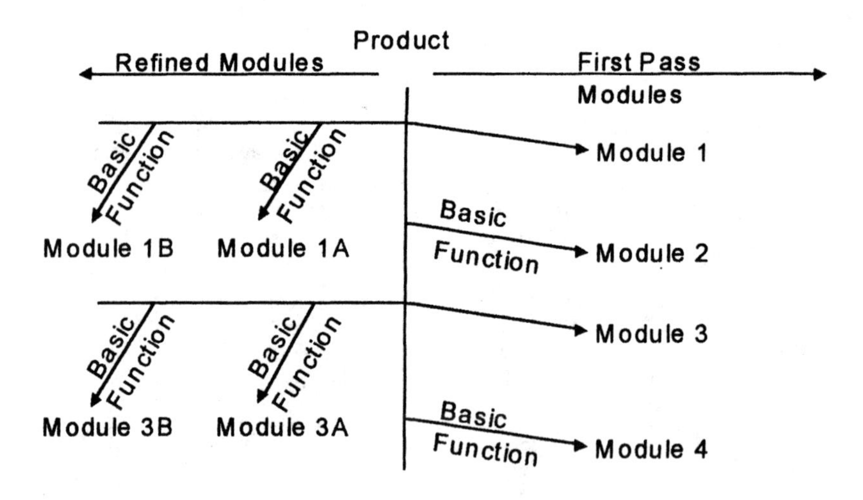

Figure 4.7 The module decomposition diagram [Reprinted with permission from K. Allen].

Table 4.4 Documentation of connection parts [Reprinted with permission from K. Allen].

Module	Connection Part(s)	Connected Module
Module 2	4 Screws	Module 1A
Module 4	1 Hose	Module 3A
Module 2	1 Wire	Module 3B

Next, the function and sub-team structure of the current product are documented using a product function diagram, as shown in Figure 4.8. This diagram relates the identified modules to the team structure of the company. The first block on the left represents the product development team responsible for developing the product, in this case, the cross-functional product development team. The next block, moving from left to right, identifies the overall function of the product. The next blocks represent the sub-teams, and to their right is their as-

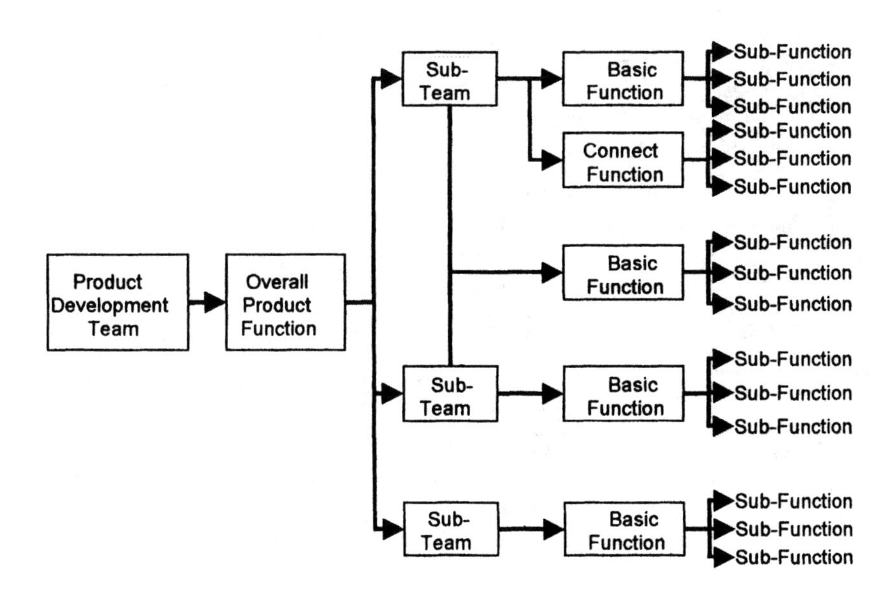

Figure 4.8 The product function diagram [Reprinted with permission from K. Allen].

signed modules that were identified in the last step. In general, most modules are assigned to one sub-team. However, if a module is the responsibility of multiple teams, then it must be coupled to these teams as shown by the lines linking them to these sub-teams. Also, connection functions are assigned to the appropriate sub-team for development. Finally, all sub-functions are identified and associated with the appropriate modules. The team structures used in small companies differ from those in large companies in that they involve fewer people. In general, the sub-teams represented in the diagram above could consist of only one person or possibly a group of people if the sub-team is responsible for multiple modules.

It is important that all of the members of the cross-functional product development team are present for this step in the process to ensure that the modules are assigned to the appropriate sub-teams. Furthermore, if this product is an extension of a product line that has similar functions, then it makes sense to assign similar modules across product lines to the same sub-teams. In addition, the team may decide to use a module "as-is" from another product in this new product. The CDT should be present for these important decisions. At the end of this step in the process, the team should know which sub-teams are assigned to which modules, and the basic functions and sub-functions that these modules provide.

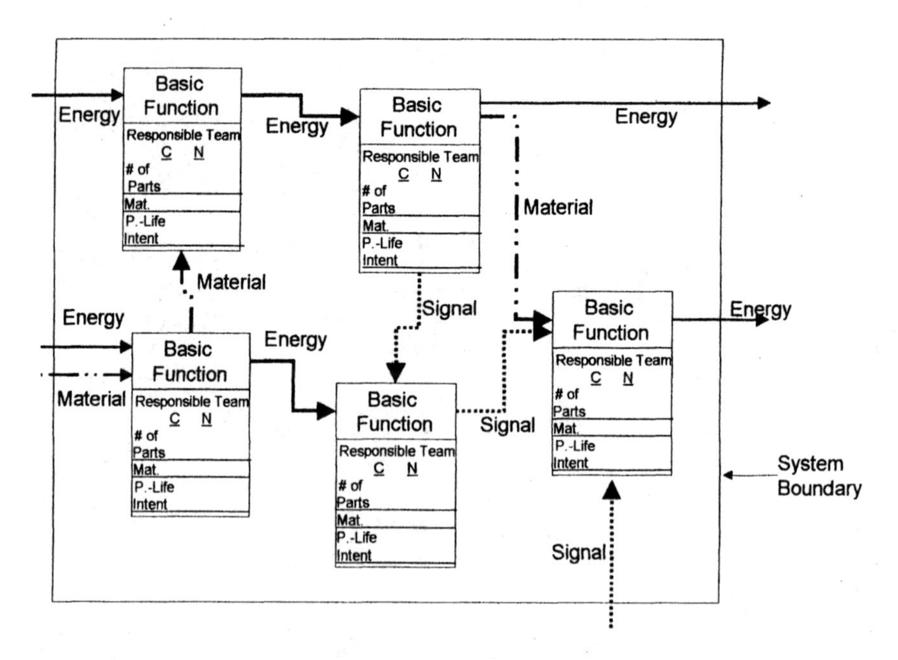

Figure 4.9 The system function diagram [Reprinted with permission from K. Allen].

The third step in defining product architecture is to define the system function structure of the current product. A diagram will be used to identify the flow of energy, materials, and signals between the modules. This important diagram identifies what information is needed by a module's sub-team from other sub-teams. This diagram is shown in Figure 4.9 and is based on a system function diagram [Pahl and Beitz, 1996].

All physical interactions are represented by flows between modules. The connection modules shown in the product function diagram in Figure 4.8, are represented as flows. The function of connection parts and modules are to join modules together. For example, if a module is connected to another with four screws, then there are four flows of translational energy between these modules. If one module produces heat and is located closely enough to another module to impact it, then a flow of thermal energy exists between these two modules.

In the system function diagram shown above, each module has associated data that documents information about the current and the new product. Each block names the responsible team as assigned in the previous step. Next there is information to help the team in improving the cost and environmental impact of the product, including number of parts, material compatibility, and the post-life intent. The number of parts can impact the assembly time and is simply the total number of parts in the module. Material compatibility gives an indication of the possibility of recycling and is calculated using the following formula:

$$Material\ \ compatibility = \frac{N_{Materials}}{N_{Parts}} \tag{4.1}$$

and is basically the number of materials in a module divided by the number of parts in the module [Allen, 1998]. The post-life intent indicates what happens to a module at the end of the product's life. This indicator is given by a number of 1 to 4 as indicated below:

1. Reuse the module "as-is" for the same or a similar application.
2. Recycle the module without further disassembly.
3. Disassemble the module further with each disassembled part being either recycled, reused, or disposed.
4. Dispose of module.

As indicated above, this information is given for both the current product as well as the new product, thereby showing any changes in the design, i.e., improvements or degradations. The numbers for the new product will most likely change during the product development process, therefore this diagram should be updated appropriately.

The final step in defining the architecture of the new product is to calculate the modularity metrics. The first of these metrics measures modularity, and the second indicates inter-module interactions. The modularity measure [Allen, 1998] shows the degree to which the product is composed of modules, and is calculated as follows:

$$Modularity = \frac{\sum_{1}^{N_{Modules}} \frac{1}{N_{Parts}}}{N_{Modules}} \tag{4.2}$$

The numerator is the sum of the inverse of the number of parts for each module. The range of the numerator is from 0 to 1. A product having a modularity close to zero indicates a product of higher modularity as opposed to one having many single-part modules. This measure can be improved by increasing the number of parts. However, this is not always a good solution as there is a tradeoff in that adding parts increases complexity, cost, and assembly time. The designers must balance between modularity and ease of assembly. Design for assembly will not be addressed in this book, and the reader is advised to refer to texts on that subject [Boothroyd, Dewhurst, and Knight, 1994]. The interactions metric [Allen, 1998] is a measure of the number of interactions between modules, and is calculated as follows:

$$Interactions = \frac{N_{Inter-Modulae\ \text{int}\,eractions}}{N_{Modules}} \tag{4.3}$$

The number of inter-module interactions are those that occur within the system boundaries. The minimum number of this measure is one, indicating a single module product. There is no maximum value; however, from empirical data gathered from disassembling a number of product, a value of three indicates problems with too many interactions and points the designers to interaction reduction.

As with all steps in the concurrent engineering process, iteration is to be expected and encouraged. Iterating through these steps helps ensure data accuracy. Finally, the team should review the results and discuss how the product can be improved.

As with the outputs in the previous design steps, the output should be documented in the same form as the others for consistency. Table 4.5 shows the information required for this design step.

4.2.2 Generate Concepts

At this point in the process, the management team should have approved the project. Now the sub-teams, with assistance from the new technology team and possibly suppliers, can begin generating concepts for the functions developed in the last design step. During this step, the teams should consider generating multiple concepts for each function, using off-the-shelf components when feasible, the modularity metrics and assembly issues discussed in the last section, the disposal, reuse, or recycling of the product, and with an understanding of how the concepts can be integrated into a functional and aesthetic design.

Table 4.5 Output for the design step: define architecture/functions and assign sub-teams.

Title	Product Structure
Definers: Primary	New technology team, Cross-functional product development team
Definers: Support	
Overview	
Original Design	Product Function Diagrams
	Sub-team Assignments
Evolutionary and	Module Decomposition Diagram
Incremental Design	Product Function Diagram
	System Function Diagram
	Sub-team Assignments
	Modularity Metrics
Documentation	Rationale

In the last section, we discussed how functions tell *what* the product does. We are now concerned with *how* the product will perform these functions. Multiple ideas on how these functions will be fulfilled is important. Figure 4.10 can help guide the teams through the process of concept generation. This process starts with the review of the product specifications and the function diagrams. The product specifications are used in the selection of the best concepts to fulfill all functional needs. Concepts that do not meet product specifications should be eliminated. The function diagrams indicate how the sub-teams are related, what the inputs and outputs are for each function, and how the functions are related to each other, and what connections must be considered. In searching for solutions, the sub-teams should rely on their past experience, but should supplement this knowledge with published literature such as industrial publications, catalogs, journals, conference proceedings, new product announcements, and patents. Often outside consultants can be helpful.

When a feasible existing solution is found, then the team should list that solution and continue searching for more. If no existing solutions can be found for a function, then the team should use brainstorming techniques to generate ideas to fulfill the solution. If there are still problems in finding concepts to fulfill a particular function, then that function should be revisited to see if it needs further refining. A quick sketch of the concept is useful, especially when communicating with other team members. Once a set of concepts or solutions has been generated or found for each function, then a selection matrix [Pugh, 1981], shown in Figure 4.11, can be used to choose the best two or three concepts for further consideration and development. In this figure, on the top left corner, the sub-team or team enters their team name, the members attending the meeting, and the date the selection matrix is created. To the right, the topic and different solutions or concepts are entered. Next, the selection criteria are defined. These criteria should be the product specifications that are relevant to that particular function, and the criteria should be designated as either a wish (W), which is desired, or a demand (D), which is required. Next, a weight factor (W_i) is defined for each criterion according to the importance of the criterion to the functionality of the product. Note that the sum of the weights must add to one.

Next, we want to evaluate the solutions or concepts. First, a benchmark must be selected, which can be an existing solution or a previous product concept. This benchmark is given a score of 0. All other solutions or concepts are rated against this benchmark; if the solution or concept is better than the benchmark, then it is given a value of +1, if the same then 0, and if worse then −1. The solutions are then scored by multiplying the ratings by the weight factor. Finally, the scores are summed and the solutions are ranked based on their scores. Figure 4.11 shows the benchmark, two concepts, their scores, and their rankings.

The next step in Figure 4.10 is to compare solutions with other sub-teams to ensure that the solutions can be generated and are compatible from the point of view of inputs and outputs. If there are problems with the concepts, then the teams should work together to generate alternative solutions or concepts for their functions. However, conflicts can be avoided if communication occurs regularly between the sub-teams and team. Finally, a set of two or three of the promising concepts for the functions are selected and documented with a sketch and short description of each. Again, as in each of the other design steps, an output should be generated. Table 4.6 shows the output for this design step.

4.2.3 Virtual/Physical Modeling

To help evaluate the best concepts and to ensure that all team members have a thorough understanding of the concepts, virtual and/or physical modeling of the solutions should be done. These models can range from computer models or virtual reality models to physical models made from foam or plastic and constructed by hand or by rapid prototyping machines. Since these models are for proof-of-concept, they need not be exact models of the product, but should represent the key characteristics of that concept so that they can be evaluated against others under consideration. It is in this step that the team is able to gain a thorough understanding of the features and functions of the new product. The models may also highlight problems with the concepts, unforeseen in paper models and that may not have been taken into consideration earlier in the design process. Table 4.7 shows the output from this step.

4.2.4 Evaluate Concepts

As in the concept generation step, this step uses the selection matrix shown in Figure 4.11 to evaluate those concepts selected for modeling in the previous step. In addition to the CDT, the customer may be asked to evaluate some of the concepts.

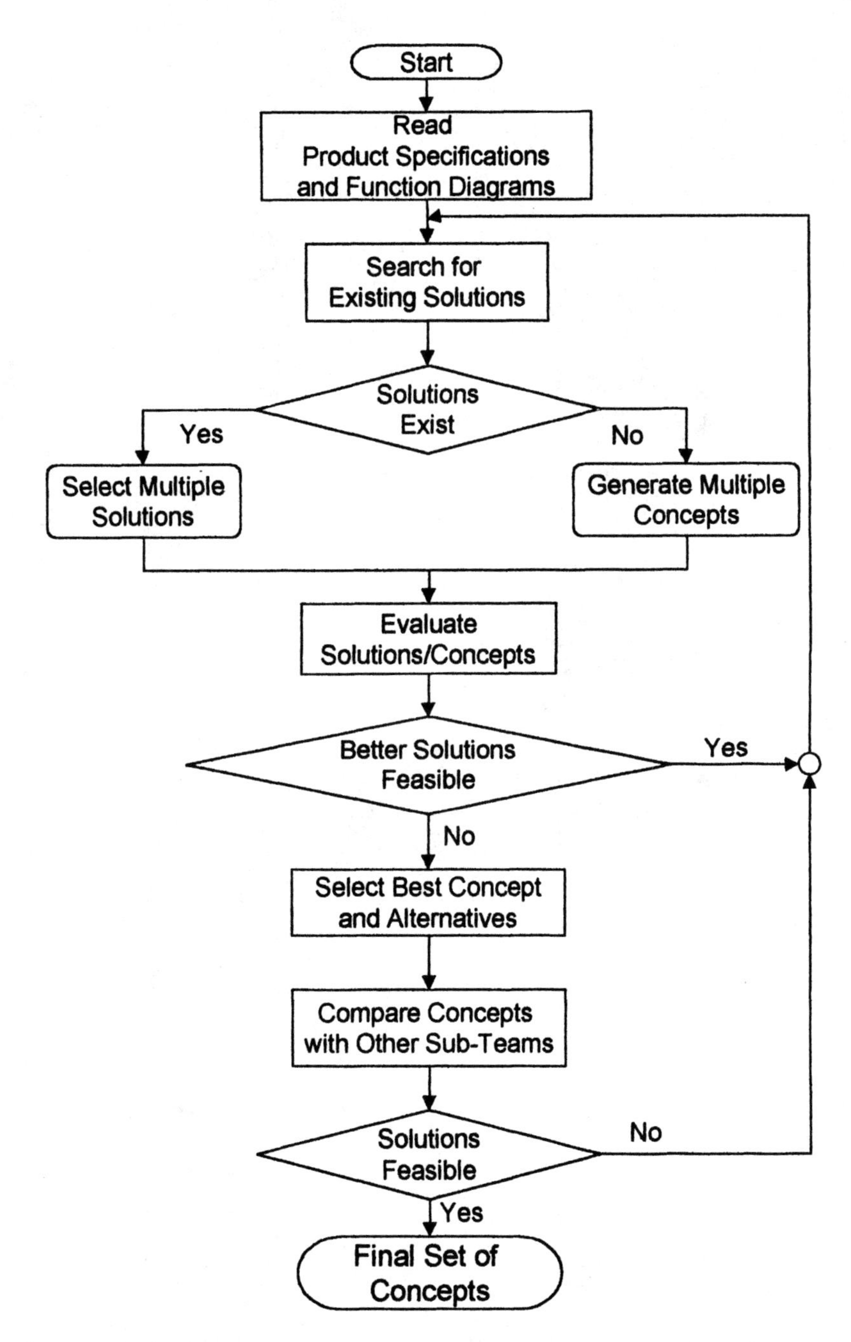

Figure 4.10 Concept generation process [Reprinted with permission from H. Kemser].

	Team:							
	Members:							
	Date		Benchmark		Concept A		Concept B	
W/D	Evaluation Criteria	Weight (0...1)	Rating (-1,0,+1)	Score (RiXWi)	Rating (-1,0,+1)	Score (RiXWi)	Rating (-1,0,+1)	Score (RiXWi)
W	Ease of use	0.4	0	0	1	0.4	-1	-0.4
W	Energy use	0.3	0	0	0	0	0	0
W	Low cost	0.3	0	0	1	0.3	0	0
	Total	1	0	0		0.7		-0.4
	Ranking			2		1		3

Figure 4.11 Concept evaluation matrix [adapted from Ulrich, Karl T. and Steven D. Eppinger, *Product Design and Development*, McGraw-Hill, Inc., New York, 1995 and used with permission from McGraw-Hill, Inc.]

Table 4.6 Output for the design step: generate concepts.

Title	Product concepts
Definers: Primary	Cross-functional product development team
Definers: Support	New technology team, suppliers
Overview	Sketches and descriptions of selected concepts
Documentation	Sketches of generated solutions and concepts
	Concept evaluation matrix

Table 4.7 Output for the design step: virtual/physical modeling.

Title	Product Models
Definers: Primary	Cross-functional product development team
Definers: Support	
Overview	Sketches, working models, computer models or other generated representations of key product concepts
Documentation	

Involving the customer early in the process can help keep the team on track in providing a product that will fulfill customer needs and that will more likely be successful in the marketplace. In this step, the team may find that it can produce a better product by combining several features of two different concepts, thereby generating another option. Then, the team must iterate through the design steps to develop this new concept's model and evaluate it against the others. As with the spiral discussed in the previous chapter, with each step and iteration, the team moves ever closer to the center and a final set of concepts. Table 4.8 shows the output for this step.

4.2.5 Integrate Concepts

In this step, the concepts of the sub-teams are integrated into an overall product concept. If the teams communicated throughout the conceptual design phase, this step should be fairly straightforward and require minimal changes and should not require the generation of new concepts for any of the functions. The output of this step is shown in Table 4.9, and should be a set of final concepts for the new product. These concepts should be described in detail and include a reason for their selection over other designs.

The conceptual design phase is a time consuming process of concurrent engineering. A considerable effort must be spent in defining product specifications, searching for concepts and solutions, and modeling and evaluating concepts. However, this time is well spent, as changes later in the process are minimized, leading to reduced development time and costs. Furthermore, using concepts and solutions from previous projects can also reduce time and effort. These benefits can only be realized, however, if all team members generate the output required at the end of each design step. The initial effort in this process can greatly reduce the work in succeeding projects.

As a final part of this design phase, the CDT must present the management team with their integrated concepts for the new product in the concepts approval milestone. The major purpose of this milestone is to ensure that the best concepts have been selected for the product, and that they integrate well with each other and with other products within the company. In the next phase, these concepts will take on a physical form.

Table 4.8 Output for the design step: evaluate concepts.

Title	Selected Concepts
Definers: Primary	Cross-functional product development team
Definers: Support	Customer
Overview	Sketches, descriptions, or models of selected concepts
Documentation	Concept evaluation matrix

Table 4.9 Output for the design step: integrate concepts.

Title	Final concepts
Definers: Primary	Cross-functional product development team
Definers: Support	Customer
Overview	Sketches, descriptions, and models of integrated concepts
Documentation	Rationale for selection of concepts and rejection of others

4.3 THE DESIGN PHASE

The design phase is the core of the CE development methodology, in which the concepts of the previous phase are given physical embodiment. In this phase, the design will move from the qualitative to the quantitative, and iteration is to be expected. This design phase, shown in Figure 4.12, consists of seven design steps: define engineering specifications, embodiment design, virtual modeling, design review, prototyping, detail design, and design verification. As in the previous sections, each design step and their outputs will be discussed in more detail. This design phase concludes with the design approval milestone.

4.3.1 Define Engineering Specifications

In this design step, the product specifications developed in the last phase will be used as a starting point for defining the engineering specifications. This step's oval shape indicates that communication is key, and the CDT should ensure that it communicates with the other teams involved in the design, especially the cross-functional process development team. Any product specifications without associated values must now be defined to include target values. Furthermore, since we now have selected concepts for the design, specifications pertinent to those concepts should be developed with their associated target values. Table 4.10 shows the output for this design step. The checklist [Pahl and Beitz, 1996] in the overview section should help the team define the engineering specifications more completely.

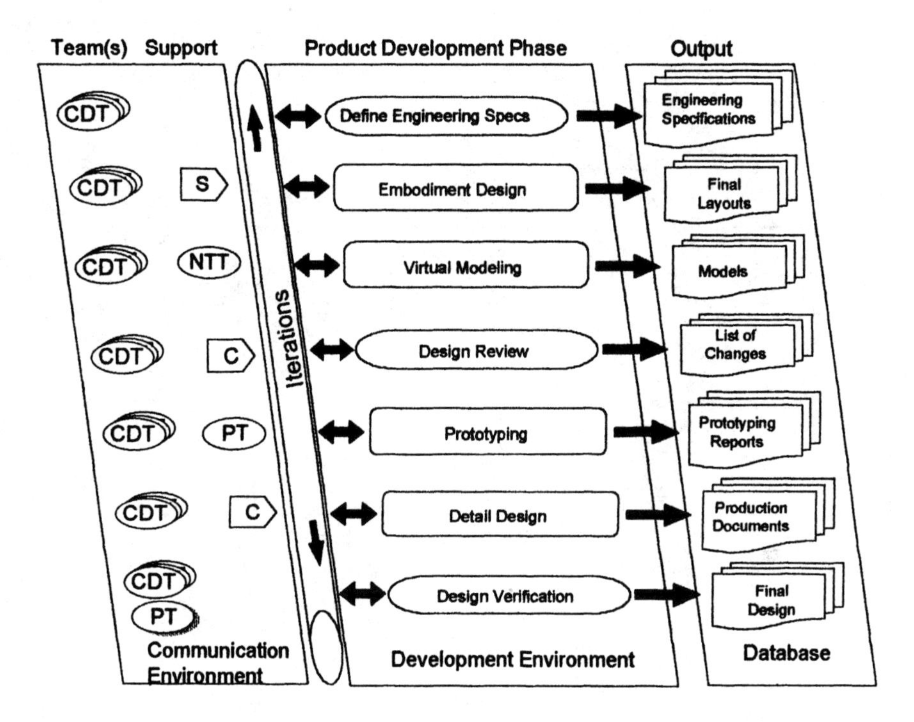

Figure 4.12 The Design Phase and its design steps [Reprinted with permission from H. Kemser].

Table 4.10 Output for the design step: define engineering specifications [Adapted with permission from Springer-Verlag GmbH & Co.KG from G. Pahl and W. Beitz, *Engineering Design: A Systematic Approach*, Springer-Verlag, New York, 1996, Figure 5.7, p. 133].

Title	Engineering Specifications
Definers: Primary	Cross-functional product development team
Definers: Support	
Overview: Specifications	
Size Requirements	Footprint
	Component sizes
	Connection sizes
	Output
Layout Requirements	Flow
	Motion
	Position
Material Requirements	Strength
	Impact resistance
	Corrosion resistance
	Heat resistance
	Deformation
	Expansion
	Durability
	Life
	Cost
	Recyclability
Safety	Components
	Operation
	Environment
Ergonomics	User interface
	Repetitive motion
Assembly	Time
	Skills
	Equipment
Transport	Vibration resistance
	Weather related issues: heat, humidity, etc.
Operation	Conditions
Environment	Disassembly, remanufacture, recycling
Maintenance	Replacement parts
	Service
Documentation	Rationale for target values

4.3.2 Embodiment Design

Embodiment design is undertaken with the support of the suppliers and seeks to give physical meaning to the concepts generated in the last design phase. In this book, we will only briefly describe this design step. Several books cover this step in detail and are listed at the end of this chapter. Embodiment design depends on the design type undertaken, however, Figure 4.13 shows a general process that can be followed for this design step. Communication is key to this design step, and the team members should stay up-to-date with decisions among themselves and communicate regularly with the process design team. As shown in Figure 4.13, the designers should first start with the main functions of the design, then move to the sub-functions. For these functions they first should consider using off-the-shelf components, and if none can be found to fulfill the requirements, then they should design that component. The output from this step should be a final layout. Table 4.11 shows the output requirements for this step.

Table 4.11 Output for the design step: embodiment design.

Title	Final layouts
Definers: Primary	Cross-functional product development team
Definers: Support	Suppliers
Overview	Final layouts
Documentation	Preliminary layouts General arrangement Components Shapes Materials
	Detailed layouts Major components and relationships Components Calculations Standards and regulations

4.3.3 Virtual Modeling

Modeling is used in this design phase as method to virtually test the design or, in other words, use a mathematical approximation to analyze its performance. This design has now moved from a concept into a more detailed form, although not yet a physical product; there are enough details from the previous design step to define computer models of the product. This step is especially straightforward if the detailed layouts are in a CAD format. There are many programs now

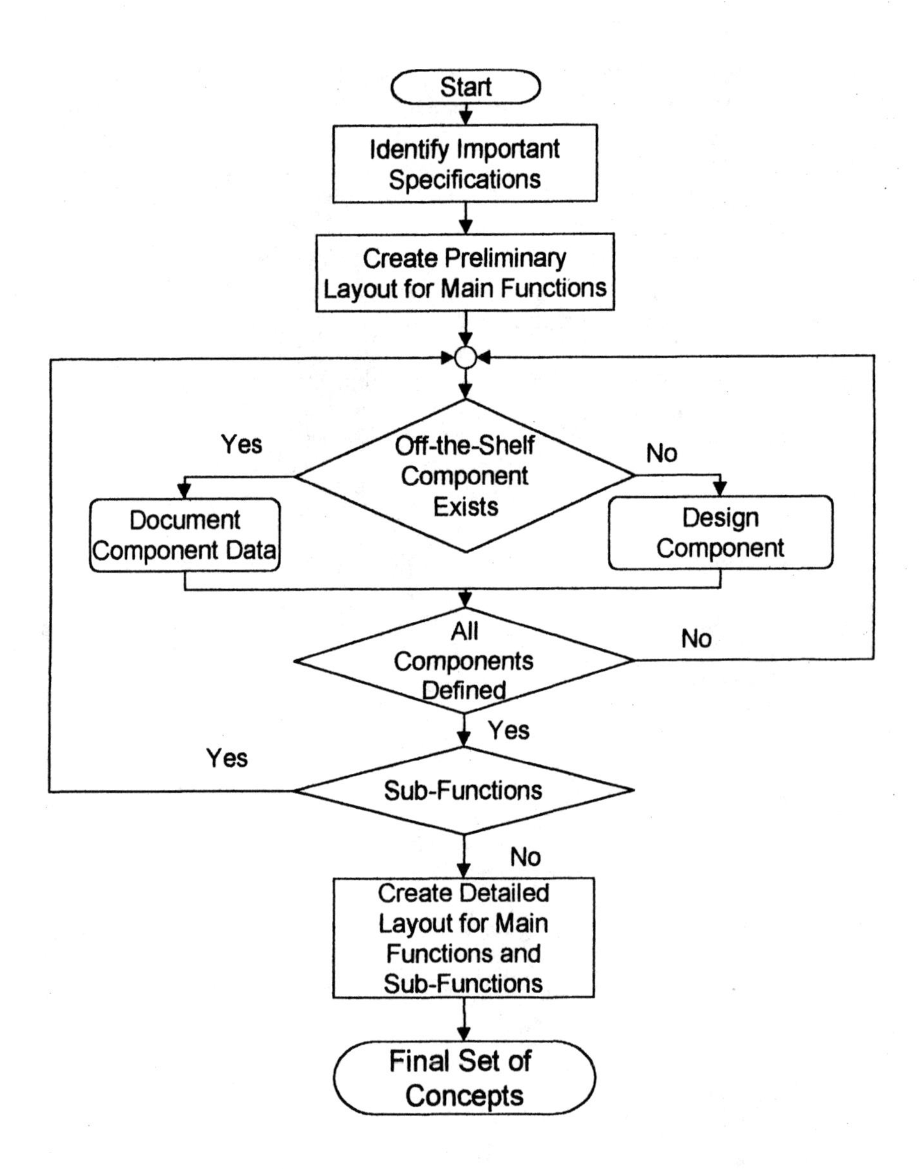

Figure 4.13 The process of embodiment design step [Adapted with permission from Springer-Verlag GmbH & Co.KG from G. Pahl and W. Beitz, *Engineering Design: A Systematic Approach*, Springer-Verlag, New York, 1996, Figure 7.1, p. 201].

available to take CAD formats and perform analysis on them, such as finite element analysis, mold-fill analysis, and interference modeling. Not all the models need be computer models as quick mathematical approximations can also be useful. This step can be very time consuming, so it is important to clearly identify what information can be gathered from which type of model (mathematical, finite element, manufacturing process model, etc.). Recall that the end goal is to develop a product that meets customer and engineering requirements. Therefore, the team must balance between the cost and time to develop the models and the desire to provide a high quality product, and meeting the delivery deadlines. Table 4.12 shows the output needed for this design step.

Table 4.12 Output for the design step: virtual modeling.

Title	Models
Definers: Primary	Cross-functional product development team
Definers: Support	
Overview	Brief description of models developed and findings from these models
Documentation	Rationale for each model developed Information desired from developed models
	Design models that can take on many forms Finite element models Engineering analysis and mathematical models Simulation Rapid prototypes
	Findings from each model

4.3.4 Design Review

This step is one of the most critical in this phase. In this step, the CDT and CDPT should have a formal meeting in which all designed components, layouts, and model results are reviewed. The main purpose of this review is to ensure that the design can be integrated into an overall functional design, and that the design can be manufactured in the facilities available. Often, the teams find missing information or data that requires an iteration back to a previous design step. This key design step is often overlooked in a company and is done on an informal basis if at all. This step is well worth the effort and time that it requires of the team members. Often implicit desires of the team members become known as they see how

Table 4.13 Output for the design step: design review.

Title	List of changes
Definers: Primary	Cross-functional product development team
Definers: Support	Cross-functional process development team, customer, field service
Overview	List of changes to be made to meet specifications and to accommodate design integration, manufacture, test, packaging, and field service
Documentation	Evaluation results and rationale behind suggested changes
	Parts and documents affected by changes

their part of the design fits with the other designed parts of the product. The main outputs of this step are changes to be made to accommodate design integration or manufacture. Table 4.13 shows the output for this step.

4.3.5 Prototyping

In this design step, physical prototypes are created for the critical aspects of the product. These prototypes can be used to test the functionality of the product or its appearance. The designers can decide to create a prototype that illustrates only one or a few functions. For example, if designing a printer, a designer may decide that he needs a prototype to test the paper feed mechanisms. This prototype would not print on the paper, but only feed it, ensuring that one sheet at a time is fed and that the design is "jam free". A fully operational prototype can also be built that illustrates a fully functional product; these are called beta products. These are often the prototypes shown to customers. There are three main purposes to developing prototypes [Ulrich and Eppinger, 1995]:

- Understanding: The development team needs to ensure that their design works as intended and that it meet customer and engineering requirements. They can also be used to ensure that the design's subsystems integrate and function as intended.
- Communication: Prototypes enhance communication between team members, other teams, and with customers, suppliers, and field service.
- Milestones: A prototype can assure the team and management that the product's design is operational. Often these prototypes are used for testing the durability and life of the product.

The danger in this design step is to try and prototype every function. The team must carefully consider what functions need prototyping and what tests will be performed on the prototypes. Building a fully functional prototype may not be

Table 4.14 Output for the design step: prototyping.

Title	Prototyping reports
Definers: Primary	Cross-functional product development team
Definers: Support	Customer, field service, supplier
Overview	Results from prototype tests
Documentation	Prototypes
	Test plan and rationale for prototypes and tests
	Results of tests

justified, so the team must carefully balance between building prototypes for testing those functions that need it against scheduling and cost demands. Table 4.14 shows the output for this design step.

4.3.6 Detail Design

In this design step, all detail and assembly drawings should be completed along with a bill of materials. In the detail drawings, the dimensions should be toleranced, and part materials and manufacturing instructions should be completed. In the assembly drawings, the drawing should show how the parts fit together, and each part should have a number that corresponds with an entry on the bill of materials. In general, most companies have their own standards for their drawings. These standards should be outlined here for the team members to follow. Table 4.15 shows the output for this design step. Notice that the product team is included in this design step, as they are the team that sees the product through production and is responsible for supporting it throughout its lifecycle.

Table 4.15 Output for the design step: detail design.

Title	Production documents
Definers: Primary	Cross-functional product development team
Definers: Support	Product team, customer
Overview	Detail drawings
	Assembly drawings
	Bill of materials
Documentation	

4.3.7 Design Verification

The goal of this design step is to check all drawings and documentation for completeness and for sign-off on the drawings by the appropriate parties. This step should be accomplished through a series of meetings with the appropriate signatories and those responsible for the drawings. Table 4.16 shows the output for this design step. If the drawings must be changed before acceptance, these changes must be noted and justified in the documentation. This is the last step of the design process before production and is therefore key to finalizing the design.

Table 4.16 Output for the design step: design verification.

Title	Final Design
Definers: Primary	Cross-functional product development team, Product Team
Definers: Support	
Overview	Detail Drawings with approval signatures
	Assembly Drawings with approval signatures
	Bill of Materials with approval signatures
Documentation	Documentation for any changes needed to the drawings before approval can be made
	Rationale for changes to drawings

At the end of this design stage is the design approval milestone, where the management team must approve the product design. A document should be prepared that includes the engineering specifications, the results of the virtual modeling, the prototyping reports, and the signed and verified production documents. This documentation should be quickly gathered because of the completion of the outputs along the way. The purpose of this milestone is to ensure that the product design is complete and that it is ready to be handed over to the product team.

4.4 THE PRODUCTION PREPARATION PHASE

This last phase of the product development model is used to ensure that the product and the process can be integrated, and the product is ready for production [Clausing, 1994]. This phase consists of four design steps: product procurement, field trials, pilot production, and production validation. This phase is managed by the product team with support from the cross-functional product development team. The phase ends with the production approval milestone. Figure 4.14 shows the production preparation phase and its design steps and outputs.

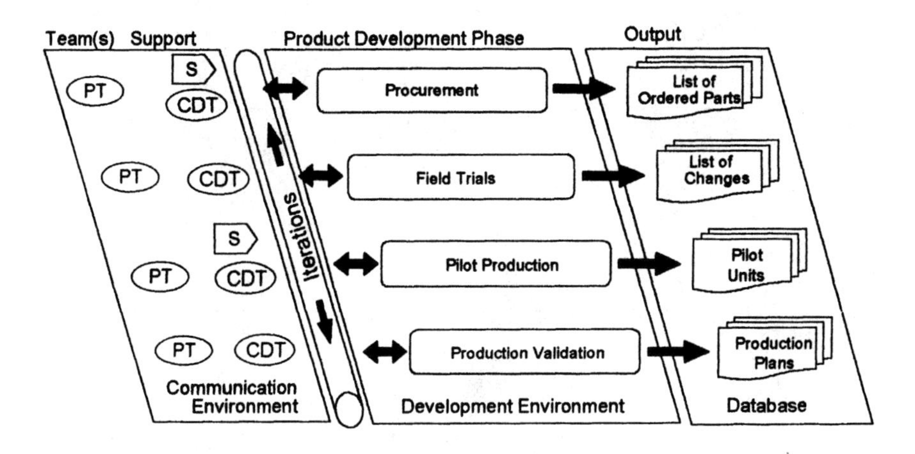

Figure 4.14 The production preparation phase [Reprinted with permission from H. Kemser].

4.4.1 Procurement

The procurement design step is the culmination of earlier work in which long lead-time parts are brought in to the facility from the vendor. Some of these parts may have been ordered in the embodiment design step in the design phase. In addition to these long lead-time parts, off-the-shelf parts that have been chosen as components in the design are also brought in for final testing and measurement. This step should be fairly straightforward since all the information required in this design step should have been considered in earlier design steps. This design step is used to ensure that all parts have been considered, and that all supplier information is up to date. Table 4.17 shows the output from this step.

4.4.2 Field Trials

Field trials are based on the product and might not always be needed for a particular project. However, in this chapter, we are showing every design step, and in Chapter 6 we will discuss when and how this step may be eliminated for a particular project. In general, if a product is to be made in high volume, then field trials should be performed. Usually, alpha and beta field trials occur during this step. Alpha field trials are used to ensure that the integration of the product results in a functional product that meets the product and engineering specifications. Normally, the alpha field trial parts are not manufactured on production ma-

Table 4.17 Output for the design step: procurement [Adapted with permission from C. Henson].

Title	List of ordered parts
Definers: Primary	Product team
Definers: Support	Cross-functional product development team, suppliers
Overview	Part name
	Supplier
	Part cost
Documentation	Assembly drawings containing parts
	Part number
	Lead time for each part
	Rationale for choosing each vendor
	Supplier information: name, address, phone number and contact names
	Approved alternate suppliers Part cost information Contact information Lead time
	Suppliers of parts that were not approved

chines, but are similar in form and material to the final parts. Alpha trials are generally conducted in-house or may be conducted at selected customer sites. Beta field trials are used to test the product in a full-use environment and should uncover any problems that arise during operation on a day-to-day basis. These trials can also be used to test customer satisfaction with the new product. Changes required at this stage of the design process should be minimal.. The output of this step will be any changes that must be made to the product. The output is shown in Table 4.18.

4.4.3 Pilot Production

As in the field trials, pilot production may not occur on all products; it depends on the volume of products produced. However, if the volume is high (>100), then it is recommended that the company produce products on the full production line. The goal of this effort is to ensure that all problems on the line are worked out before production begins. These products will not be sent to customers, but should be examined and tested for compliance with all customer and engineering requirements. These products may then be used in-house. In addition to product assurance, production personnel, the production process, and production plans are also put to the test to ensure that all goes as planned. This is the design step in which any problems should be corrected before customer products are made. Table 4.19 shows the output for this design step.

Table 4.18 Output for the design step: field trials [Adapted with permission from C. Henson].

Title	List of changes
Definers: Primary	Product team
Definers: Support	Cross-functional product development team, customer
Overview:	Changes listed in order of occurrence
	Engineering change number for tracking
	Change approval
Documentation	Benefit of change
	Who prompted change (customer, internal, ...)
	Responsible person
	Affect on project: schedule and cost
	Ability to satisfy customer needs

4.4.4 Production Validation

This is the final design step before full scale production occurs, which will be discussed in the next chapter. This step is used to ensure that all production processes are working as intended. The production rate should be checked and verified, and production quality should be monitored. We have not discussed product quality, but key quality measurements should be monitored to ensure that the process maintains the product quality intended. See the References and Bibliography list for books that cover these issues in detail. In this production run, each section of the production line should be monitored and the results of the run should be recorded as indicated in the output in Table 4.20.

At the end of this design step, the customer/production approval milestone takes place. At this milestone, the management team needs to guarantee that all parts are ready and available for production, that all changes needed to the product are completed, and that all production documents are available. Again, as in the previous milestone, this documentation should be easily gathered from the outputs of the design steps. If the concurrent engineering process and teamwork have been adhered to, this step should be a formality. This milestone concludes the product development model.

Table 4.19 Output for the design step: pilot production [Adapted with permission from C. Henson].

Title	Pilot units and results
Definers: Primary	Product team
Definers: Support	Cross-functional product development team, supplier
Overview:	Resulting pilot products
	List of problems encountered and solutions proposed and undertaken
	Production rate
Documentation	List of problems Proposed solution Personnel responsible Schedule for solution implementation Date of completion
	Product quality check Results Recommendations Schedule for implementation (if needed) Responsible person

Table 4.20 Output for the design step: production validation [Adapted with permission from C. Henson].

Title	Production plans
Definers: Primary	Product team
Definers: Support	Cross-functional product development team
Overview:	Report of production run results
Documentation	Separate production plan for each production area Requirements for that section of production Characteristics of the product produced in that area Tolerances to be checked Frequency of checks Inputs and output to the section Referenced documents
	Safety documents for each section • Material safety data sheets Emergency plan and documentation

4.5 SUMMARY

The product development model consists of four design phases, which include project planning, conceptual design, design, and production preparation. These development phases are carried out by primarily by the cross-functional product development team with support from the new technology team in the early design steps, field service, suppliers, customers, and in the later design steps by the product team. Each design phase follows the same general layout in which teams and support members are assigned to be responsible for various design steps. These teams then carry out the various design steps which results in decisions and outputs for each step which are recorded in a database. The decisions made within each phase begin very broad, but with time and iteration within the phase, the decisions become more detailed until a final milestone report for that phase is completed and ready for management approval. Each of the outputs has been detailed in this chapter and each of the milestones has been overviewed.

REFERENCES AND BIBLIOGRAPHY

Allen, Kevin, "Defining Product Architecture and Assigning Sub-Teams within a Concurrent Engineering Methodology for Small Companies," Masters Thesis, The University of Virginia, 1998.

Clausing, Don, *Total Quality Development*, ASME Press, New York, 1994.

Day, Ronald G., *Quality Function Deployment: Linking a Company with its Customers*, ASQC Quality Press, Milwaukee, 1993.

Henson, C., *Defining the Outputs of a Small Business Concurrent Engineering Model*, Senior Thesis, University of Virginia, 1998.

Ishii, K. and B. Lee, "Reverse Fishbone Diagram: A Tool in the Aid of Design for Product Retirement," *Proceedings of the 1996 ASME Design Engineering Technical Conferences and Computers in Engineering Conference*, paper number 96-DETC/DTM 1272, August 18-22, Irvine, CA.

Kemser, H., *Concurrent Engineering Applied to Product Development in Small Companies*, Masters Thesis, University of Virginia, 1997.Pahl, G. and W. Beitz, *Engineering Design*, Springer-Verlag, New York, 1996.

Pahl, G. and W. Beitz, *Engineering Design*, Springer-Verlag, New York, 1996.

Pugh, Stuart, *Total Design*, Addison-Wesley, Reading, MA, 1990.

Ulrich, Karl T. and Steven D. Eppinger, *Product Design and Development*, McGraw-Hill, Inc., New York, 1995.

SUGGESTED READING

Engineering Design by G. Pahl and W., Springer-Verlag, New York, 1996.: An in-depth text that covers product planning through embodiment design, with special concentration on a systematic approach to conceptual and embodiment design.

Quality Function Deployment by Ronald Day, ASQC Quality Press, Milwaukee, 1993: A very complete book on the use of the House of Quality.

Total Design by Stuart Pugh, Addison-Wesley, Reading, MA, 1990: Covers Pugh's approach to systematic design. He is known for his development of the concept selection chart that was adopted in this book.

Total Quality Development by Don Clausing, ASME Press, New York, 1994: Covers concurrent engineering from a quality perspective.

Product Design and Development by K. Ulrich and S. Eppinger, McGraw-Hill, Inc., New York, 1995: Focuses on methods, tools and techniques for design teams to use in the development of new products.

5

Process Design: Steps and Tools

In collaboration with Johanne Phair, Hans-Peter Kemser, and Chris Henson

At this point, we have discussed the product development model in depth, we will now turn to the process development model. Briefly overviewed in Chapter 3, this model consists of three different models: the manufacturing process development model, the test method development model, and the packaging development model. As in the last chapter, we will discuss all phases and their steps for these models. Many of the steps are very similar to those discussed in the last chapter, and as such will only be overviewed briefly. As seen in Figures 5.1-5.3, the three models each have five development phases: project planning, conceptual design, design, production preparation, and production/service. Between each of these five development phases, there is a management milestone, which is an approval point for the management team to assess the progress of the project. The four milestones are defined as project approval, concepts approval, process approval, and production approval. Each of the phases are broken into steps. In this chapter, we will discuss each phase of the models and each step within these phases. The process development model starts with a kick-off meeting in conjunction with the product development model. This is the same meeting that was discussed at the beginning of Chapter 4. This meeting is used to develop a broad overview of the development methodology and the expectations of the product. Next, the teams begin their own tasks, but communicate their progress regularly. Then, the cross-functional process development team's tasks will be discussed. Finally, production/service, the last phase of both the product and the process development models will be discussed. This is the phase in which the product is manufactured, tested, packaged, and distributed to customers. Within this phase's purview is the support of the product through field service and/or customer support organizations.

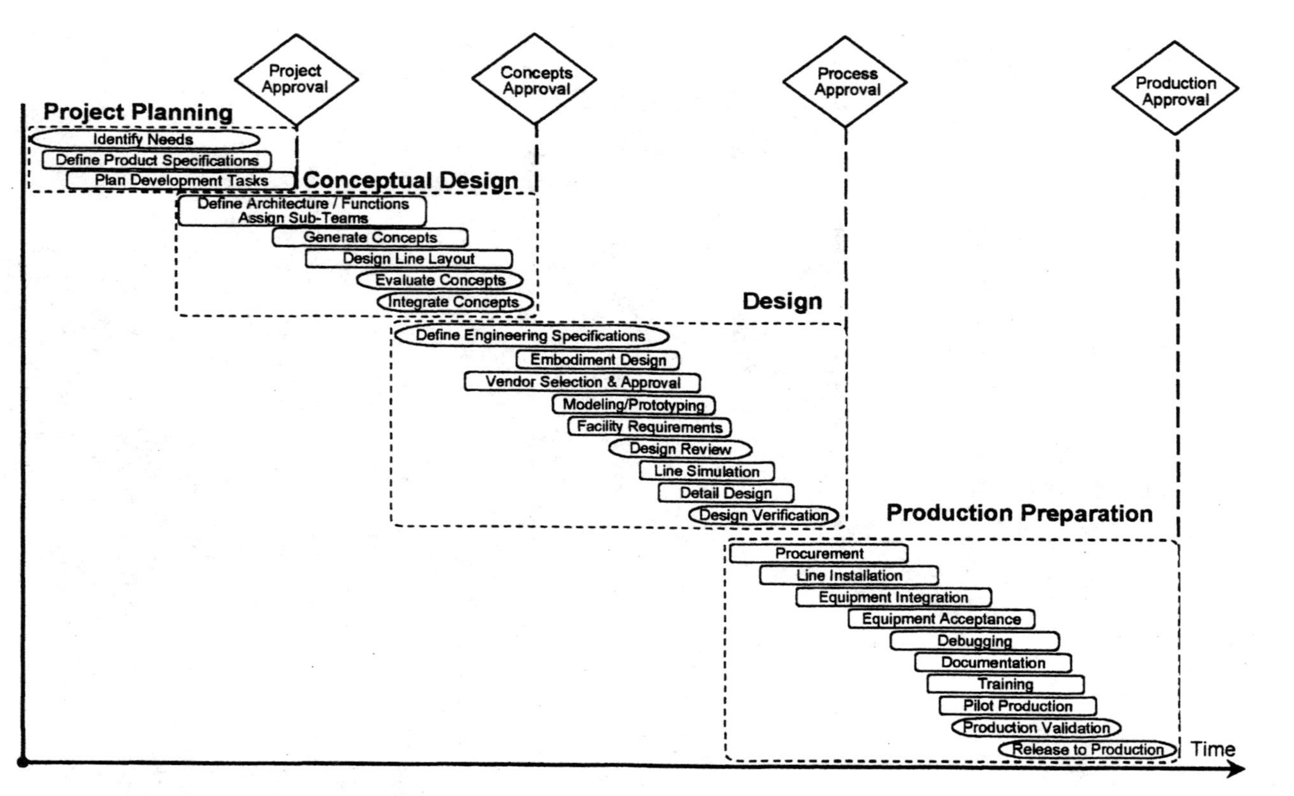

Figure 5.1 The process development model showing phases, steps, and milestones [Adapted with permission from J. Phair].

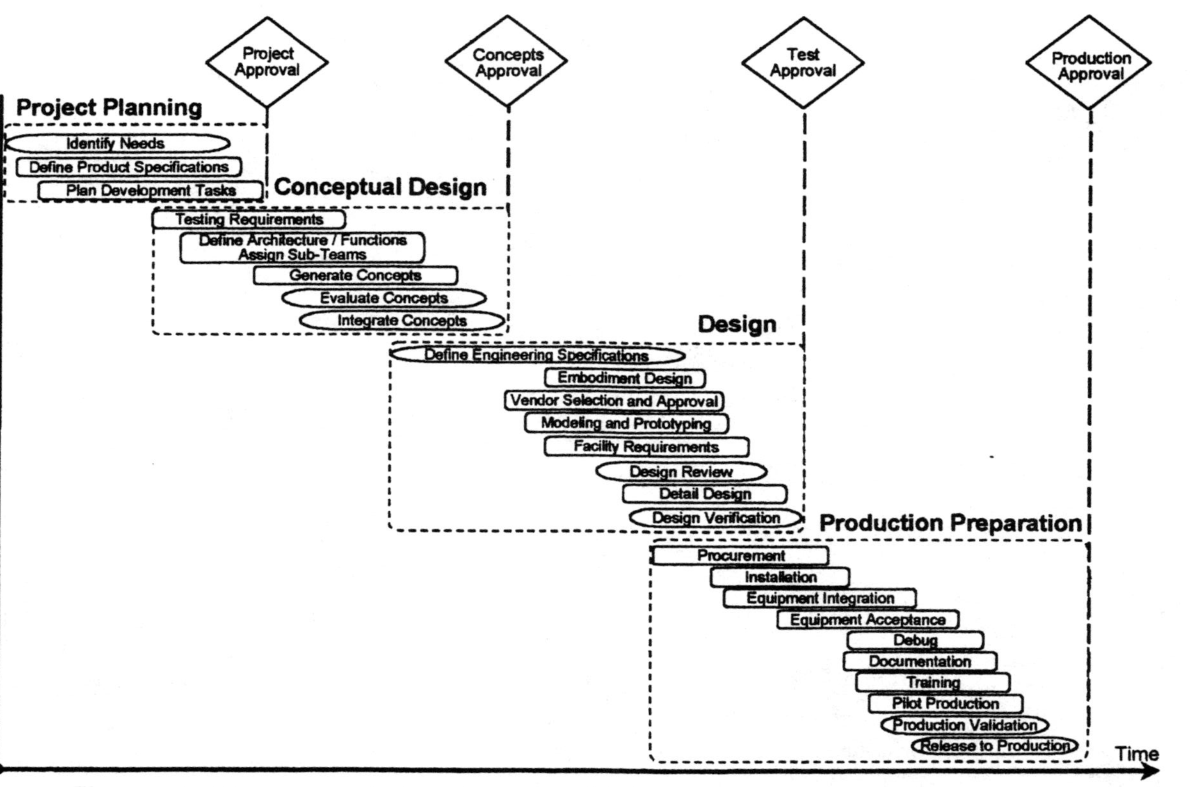

Figure 5.2 The testing method development model showing phases, steps, and milestones [Adapted with permission from J. Phair].

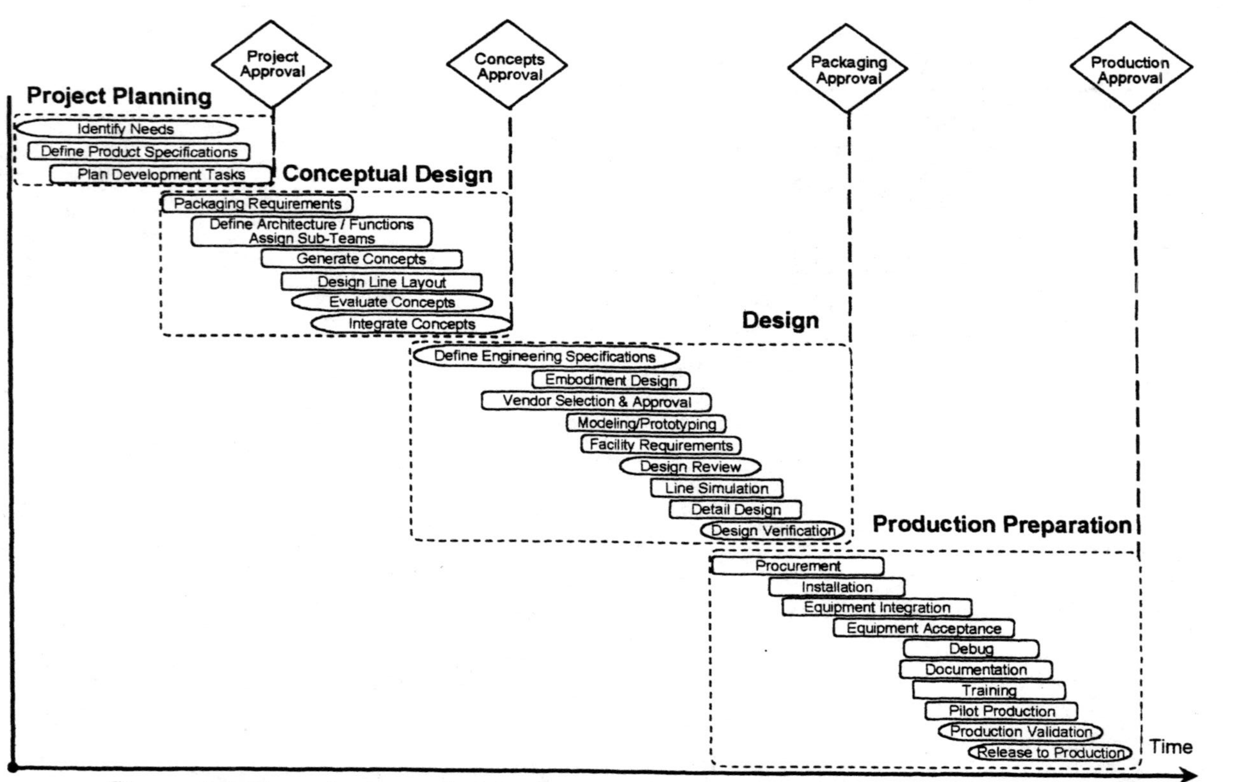

Figure 5.3 The packaging method development model showing phases, steps, and milestones [Adapted with permission from J. Phair].

5.1 THE PROJECT PLANNING PHASE

The project planning phase for the manufacturing process, testing, and packaging models are combined, as the designs from these models must be integrated and are interdependent. In this step, the designers and manufacturing personnel clearly define what the project will require from a manufacturing point of view. Figure 5.4 shows the steps of this phase. Just as in the product development model, each of the three steps of this phase has assigned teams and support personnel and a defined output. Next, we will discuss each of the steps in more detail.

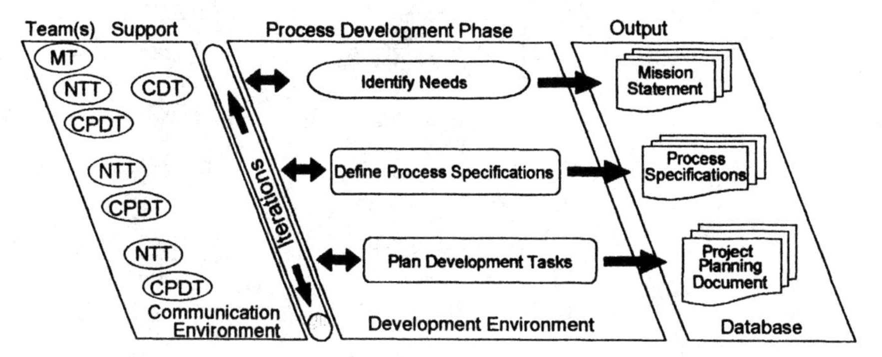

Figure 5.4 The project planning phase of the three models within the process development methodology [Adapted with permission from J. Phair].

5.1.1 Identify Needs

The first step of this phase is the most important step in the project planning phase, since the decisions made in this step guide the cross-functional process development team (CPDT) through the rest of the development process. The cross-functional product development team (CDT) plays a supportive role in this step, since they can provide information on the decisions made in the product development process that directly affect the manufacture of the product, and vice versa.

The output for this design step is titled, mission statement, although it contains much more. The outputs for each of the design steps, as discussed in Chapter 3, should have the consistent form that was used in the last chapter. There are four parts to the output: the title, the list of definers (those responsible for developing the content of the output), the overview, and the documentation. Table 5.1 defines the output for the design step: identify needs.

Table 5.1 Output for the planning step: identify needs.

Title	Mission statement	
Definers: Primary	New technology team, cross-functional process development team, management team	
Definers: Support	Cross-functional product development team	
Overview	Mission statement	
Documentation		
Company	General issues	Available resources
		Type of investment required
		Partners to provide skills or equipment not possessed by company
		In-house or purchased components
		Regulations or standards to be met
		Review past projects to look for areas of improvement, problems to avoid, and successes to repeat
	Company strategy	Type and amount of risk company is willing to undertake
		Type of risk associated with project
		Fit of project with existing product lines
		Fit with operating strategy
		Fit with vision, mission, and goals of company
	Marketing	Increased competitive ability gained through manufacturing
		Risks to company if announcement date is missed
Process needs	Design	New requirements from product designers
		New line with new or existing technology, or existing line with major or minor changes
		Special product considerations from a manufacturing point of view
	Competition	Competitive processes currently in use
	Technology	Expected functionality of processes
		Are existing technologies sufficient?
		Have new technologies been recently introduced and should they be incorporated?
		Where must machines be compatible with other machines and lines?
		Special power requirements
		Will more than one product be made on the same line?

There are many questions that must be answered in the documentation section. The first of these that will be important for the team to ascertain is the amount of design work that will be required to manufacture the new product. The level of work required has been categorized as follows: a new line with new technology, a new line using the same technology, major changes to an existing line, or minor changes to an existing line. Finally, the mission statement must be defined, which is used to guide the team through this development effort and keeps the project's aims clear and certain. By answering the questions in the documentation section, the team should be able to define a clear mission statement for this project. Also, these questions will help the designers begin to define the specifications in the next step.

5.1.2 Define Process Specifications

Process specifications broadly define what is expected from the manufacturing line and its processes. These specifications should have target values associated with them. For example, what is the needed manufacturing cycle time, and the required daily output of product from the lines? Many of these target values will be very rough, but as indicated from the diagram of the design phases, iteration is possible and expected. As more information is gathered and knowledge formed about the product and the processes needed to produce this product, the target values can become more concrete. From these process specifications, the team members can look for the means to meet the requirements. Although target values for the specifications are important, the specifications should be kept general and should not specify solutions. Table 5.2 shows the output for this design step. The checklist provided in the overview section should help the team begin to define these process specifications.

5.1.3 Plan Development Tasks

The final step in this design phase is to plan and schedule the process development tasks. In this step a rough budget should be defined for the design, modification, and build of the manufacturing lines. Also, a preliminary schedule or timetable should be defined with key time constraints outlined, which will then be expanded later by the various team members. The key questions that need to be answered in this step are:

- When is the completed manufacturing line needed?
- How can available resources be integrated and used?
- What are the unique aspects of this process development project?

Also in this step, the process development models are tailored to fit the project characteristics; this tailoring process will be defined in the next chapter.

Iteration among these three steps is expected. As new information becomes available, then revisions to the needs, product specifications, or schedule may need

Table 5.2 Output for the define planning step: process specifications.

Title	Process specifications
Definers: Primary	New technology team, cross-functional process development team
Definers: Support	
Overview: Specifications	Quantity of products needed per day
	Number of parts needed per day
	Cycle time per station
	Space requirements
	Required processes
	Process features
	Product movement
	Estimated down time
	Energy required (type and amount)
	Energy produced (type and amount)
	Waste produced (type and amount)
	Types of environment needed for processes
	Maintenance required
	End-of-life: rework, recycling, remanufacture, and/or disposal of products and waste
	Target cost of production lines
Documentation	Rationale for requirements and associated values

Table 5.3 Output for the planning step: plan development tasks.

Title	Project planning document
Definers: Primary	New technology team, cross-functional process development team
Definers: Support	
Overview:	
Estimate resources	Cost of development
	Space requirements
	Required personnel: team members and support
	Shop facilities
	Computer equipment
	Equipment needed for development and testing
	Available equipment
	Resources needed from partners or suppliers
Plan Schedule	Estimate of development and production schedule
Documentation	Rationale for resource needs and schedule

to be made. As with the product development methodology, research on new technologies is not to take place within this phase, but is done prior to the undertaking of this development cycle. If new technologies are being considered, they should be tested outside the development cycle, then brought in during subsequent steps.

As shown in the previous chapter, the key decisions made within this phase should be recorded in a database as part of the outputs. This database can be used by all team members during the project to check on progress, to get information and decisions relative to their part of the development project, and to communicate with each other. Once a project is complete, this information can be used in later projects to help speed the development of new manufacturing lines, just as it can be used in product development.

This phase concludes with the project approval milestone in which the management team will want to see the mission statement, company issues including strategy and goals for this development project, the process specifications, the budget, the required and requested resources, and the preliminary schedule. The most important question that the CPDT must answer is whether the company successfully manufacture the new product, or if not, what alternate plans have been considered. The management team will review this information and, based upon it and recommendations from the team leaders, will decide how the CPDT should proceed. As discussed in the last chapter in section 4.1.3, the team can decide to use a House of Quality (HOQ) to represent the requirements and specifications for the manufacturing lines. In this HOQ, the customer would be the CDT, since the customer does not care how the product is manufactured, only that the resulting quality is high and their requirements are met. After this planning phase, the manufacturing development method splits into its three components of manufacturing process, testing, and packaging.

5.2 THE CONCEPTUAL DESIGN PHASE

In the conceptual design phase, ideas are generated that meet the requirements outlined by the product and the process specifications. This process is much the same as what was discussed in Chapter 4 for concept generation in the product development process. First, the manufacturing process development model's conceptual design phase will be discussed. The manufacturing process development model outlines the phases and design steps for the design of the manufacturing processes needed for the fabrication of the new product. These processes are generally part of a manufacturing line or series of lines that are linked through the transportation of the product. In addition to the manufacturing process model, the testing and packaging models have their own conceptual design phases. The testing model is used to design the methods or test stations used to ensure that the product works as intended and according to the product specifications. The packaging design model is used to design the line or stations at the end of the manufacturing process line to package the product for shipment to the customer.

5.2.1 The Manufacturing Process Model

In the next several sections, we will outline each of the steps of the conceptual design phase of the manufacturing process model shown in Figure 5.5, just as we did in the previous sections. Then, we will show the testing and packaging models, which differ little from the manufacturing process model; therefore, only the steps that differ from the manufacturing process model will be discussed. However, in the next chapter, we will discuss the tailoring of each of the three models separately, as they can have very different requirements resulting in different models.

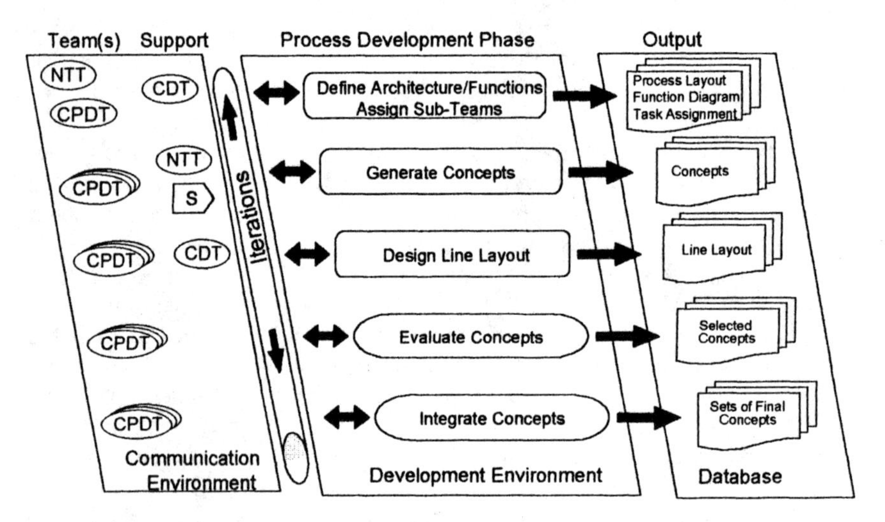

Figure 5.5 The conceptual design phase of the manufacturing process development model [Adapted with permission from J. Phair].

5.2.1.1 Define Architecture and Functions and Assign Sub-Teams

The first design step is to define the line architecture, the line functions, and assign sub-teams to the design tasks. In the planning phase, the decision should have been made as to whether the product would be made on a continuous process line or a discrete parts manufacturing line. In this book, we will concentrate on the discrete parts line, although this methodology can be applied to a continuous process line. Next, based on the product volume, the CPDT will know whether the line will be a job shop for low volumes, batch production for medium volumes, or mass production for high volumes. In this design step, the CPDT should define the line functions, that is, what tasks will be performed on the line to manufacture the product. Then, the line architecture, how the functions are laid out, can be defined. Finally, the sub-teams are assigned to the different functions of the line based on their expertise.

The CPDT will rely on the CDT as support members to help them understand the product, its functions, and requirements. The NTT is present to advise on new technologies that could prove useful for the new line. Working together, these teams should decide how the product is to be manufactured. From this understanding, the CPDT should be able to define the functions of the manufacturing line. As in Chapter 4 for product design, a function diagram is used to represent what the manufacturing line will do in the making of the new product. In this diagram, the functions are linked by the flow of material, energy and/or information. First, the overall function of the manufacturing line must be defined along with its associated inputs and outputs of materials, energy, and signals or information. Next, this overall function is broken into sub-functions. All functions are action words that define what the manufacturing line does in the making of the new product. How the product accomplishes a particular function is addressed in the next design step. The sub-functions are generally laid out in the order in which they are accomplished; this layout defines the line architecture. The sub-functions are linked by the flow of material, energy, and/or information between them. Figure 5.6 shows an example of a function diagram for the manufacture of a personal computer printer. The large, dotted-line box shows the overall function of the manufacturing line, and the smaller boxes within the overall function are the sub-functions of the line. Material flows are shown in bold, broken lines. In this case, the material is the printer as it is being built. The energy flows are shown by the solid lines and represent both potential energy (PE) and kinetic energy (KE). The signal flows are represented by dotted lines. Note that the functions are all very general and do not indicate what means will be used to install parts. These functions may need to be broken down further to allow the generation of concepts, which occurs in the next design step.

Finally, the sub-teams will be assigned to the various functions defined in the function diagrams. As discussed before, in a small company, team membership can overlap, meaning that a designer may have more than one assignment. To ensure that the line will be easy to integrate once designed, it is imperative that sub-teams communicate with each other, and that the CPDT and CDT also communicate. This communication is needed to ensure that the concepts can be integrated and that the teams are working to consistent specifications and constraints. Finally, the output for this design step should be documented. Table 5.4 shows the information required for this design step.

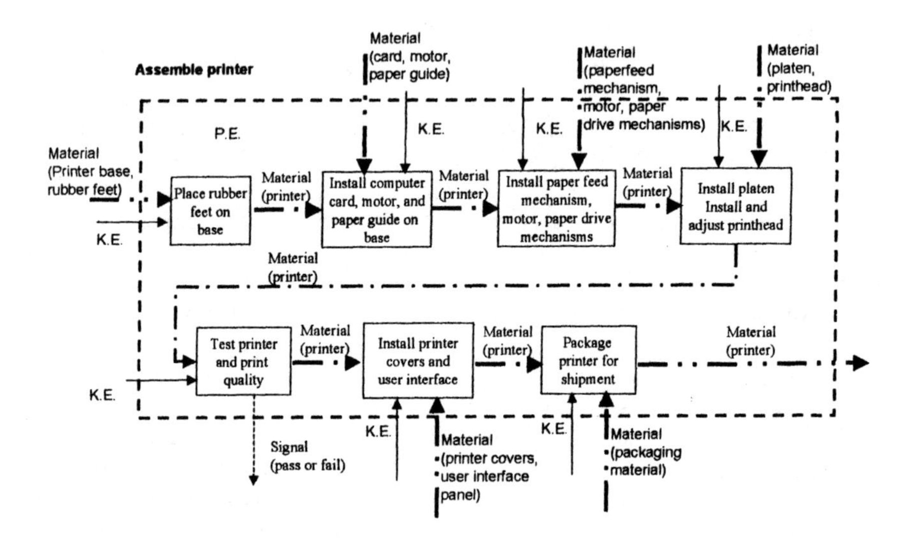

Figure 5.6 A functional diagram indicating line architecture for a personal printer assembly line.

Table 5.4 Output for the design step: define architecture/functions and assign sub-teams.

Title	Product structure
Definers: Primary	New technology team, cross-functional process development team
Definers: Support	Cross-functional product development team
Overview	Process function diagrams and line layout
	Sub-team assignments
Documentation	Rationale for functions and placement
	Sub-team tasks

5.2.1.2 Generate Concepts

The next step in this phase is to generate concepts that indicate how the functions of the manufacturing line will be fulfilled. The same principles used in Chapter 4 should be used for generating concepts for the manufacturing line. Figure 4.10 shows a general process for generating concepts. Multiple concepts for each function should be conceived, using off-the-shelf solutions when feasible. As this process moves forward, there are several questions that should be considered:

- Are there proven technologies already available?
- Can the these technologies be easily implemented to perform the specific functions needed?
- Are the technologies for the functions compatible within the same line?
- Are there technologies and machines on the existing line that can be used for the new product?
- What are other companies using to perform these functions?

The function diagram shown in Figure 5.6 should be used as a starting point. However, it is likely that the sub-teams will need to develop much more detailed diagrams like the one shown in Chapter 4 for the ice cream scoop. For example, the first function in Figure 5.6 indicates that rubber feet need to be installed on the base of the printer. The designer needs to figure out: 1) how the base arrives in the station, 2) how the feet will be presented or fed into the station, 3) how the feet will be picked up, 4) how they will be installed on the base, 5) if the base needs to be moved to allow access to all places that need feet, 6) how placement errors will be handled, and 7) how the base with its feet will leave the station to move to the next station. Designing a manufacturing line station is much line designing a product. The same process can be followed for the design of a station, and coordination between sub-teams designing different stations is key to the integration of the line. Each sub-team should generate several concepts for each task on the station. Once several concepts have been developed for each station function, then a selection matrix like that shown in Figure 4.11 can be used to choose the best two or three concepts for further consideration and development. These concepts should be compared and examined by the other sub-teams, and the support teams (in this case the NTT and the suppliers). Next, each of the chosen concepts should be documented with a sketch and short description. Finally, the output for this step (shown in Table 5.5), should be developed to document this design step.

5.2.1.3 Design Line Layout

The next step in this phase is to design the line layout. There are several factors that must be considered in this step. The line should be designed to avoid bottle necks and have a smooth flow of product. The cycle times for the stations should be balanced. The CDT should be involved to provide information on the future needs of the product and its successors. The following issues need to be considered when designing the layout:

Table 5.5 Output for the design step: generate concepts.

Title	Concepts
Definers: Primary	Cross-functional process development team
Definers: Support	New technology team, suppliers
Overview	Sketches and descriptions of selected concepts
Documentation	Sketches of generated solutions and concepts
	Concept evaluation matrix

- What are the space requirements for the line?
- What infrastructure is needed (power, water, air, ventilation)?
- What is the probability that additional capacity will be needed?
- What would be the cost of expanding capacity? Can it be easily accommodated on the line by extra stations?
- Is competition likely to affect product volume in a significant way?
- Is the technology proposed for use on the line becoming outdated? Is new technology going to be needed to replace some of the stations?
- Are follow-on products anticipated in this product line? Will follow-on products use the same manufacturing line? Should flexibility be built into the line to handle these new expected products?
- How many shifts will be needed to supply the product volume required?
- What is the expected life cycle of the new product?
- Are there enough manufacturing personnel to run the lines is a shortage of skilled labor anticipated?
- What kind of parts inventory is needed to maintain this product? Where will the parts be stored: on site, on the line, at a distribution center? How many days of parts should be stocked?
- Where will the finished products be stored?
- How will defective products be handled? Is a rework area on the line needed or will it be located elsewhere?
- How will waste be handled?
- What is the maintenance plan for the line? How much maintenance will be needed?

All of these issues must be considered when designing the line layout. As stated many times, iteration through the conceptual design phase is expected, as these are important decisions that are being made and one change can impact other parts of the design. Table 5.6 shows the output for this design step.

Table 5.6 Output for the design step: design line layout.

Title	Layout plan
Definers: Primary	Cross-functional process development team
Definers: Support	Cross-functional product development team
Overview	Line layout with description
Documentation	Sketches, working models, computer models, or other generated representations of key process concepts
	Assumptions and rationale behind data and models

5.2.1.4 Evaluate Concepts

At this point in the design process, two or three concepts for each function have been developed and selected out of a number of generated concepts. In the last design step, the line layout was developed. Now, the CPDT and the CDT must meet, as indicated by the oval shape of the design step in Figure 5.5, to discuss the chosen concepts and make a final selection. As in the concept generation step, this step uses the selection matrix shown in Figure 4.11 to evaluate the concepts. In this step, the team may find that it can produce a better product by combining several features of two different concepts, thereby generating another option. Then, the team must iterate back through the design steps to discuss the changes this new idea would make on the line layout and evaluate it against the others. Table 5.7 shows the output for this step.

5.2.1.5 Integrate Concepts

In this step, the concepts selected in the last design step are integrated. If the teams communicated throughout the conceptual design phase, this step should be fairly straightforward and require minimal changes and should not require the generation of new concepts for any of the functions. The final set of integrated concepts should be compared with the documents created in the planning phase to

Table 5.7 Output for the design step: evaluate concepts.

Title	Selected concepts
Definers: Primary	Cross-functional process development team
Definers: Support	
Overview	Sketches, descriptions, or models of selected concepts
Documentation	Concept evaluation matrices

ensure that all the process specifications have been met. As indicated by the oval shape of this design step as shown in Figure 5.5, this step should be carried out in conjunction with the CDT. This team needs to be assured that the product can be made on the proposed line in accordance with the product specifications. The output of this step is shown in Table 5.8, and should be a set of final concepts for the new line. These concepts should be described in detail and include a reason for their selection over other designs.

Table 5.8 Output for the design step: integrate concepts.

Title	Final concepts
Definers: Primary	Cross-functional process development team
Definers: Support	
Overview	Sketches, descriptions, and models of integrated concepts and final conceptual line layout
Documentation	Rationale for selection of concepts and rejection of others

5.2.2 The Testing Model

The model for the development of the testing methods and/or stations is shown in Figure 5.7. This figure differs from the manufacturing process model in three ways. First, there is a step defined as testing requirements. The next step defines the architecture and functions of the test methods. Finally, there is no step for design line layout. It is assumed that the test methods or stations will be integrated into the manufacturing process line, rather than having a stand alone line. Also notice that it is assumed there is only one team working on these design steps, which is made-up of members of the CPDT. All the steps and outputs for this model are identical to those in the manufacturing process model. Therefore, we will only discuss the step that is added, which is testing requirements.

In the testing requirements step, the team needs to determine what the critical aspects of the product are that need to be tested to ensure that it is functional and meets all product and quality specifications. It is critical that the CDT acts as a support team in this step, as they know the critical measurements, how to test the functionality, and are most familiar with the product and engineering specifications. From the requirements list, the team can then move on to find available testing means for the product. The output for this step is shown in Table 5.9.

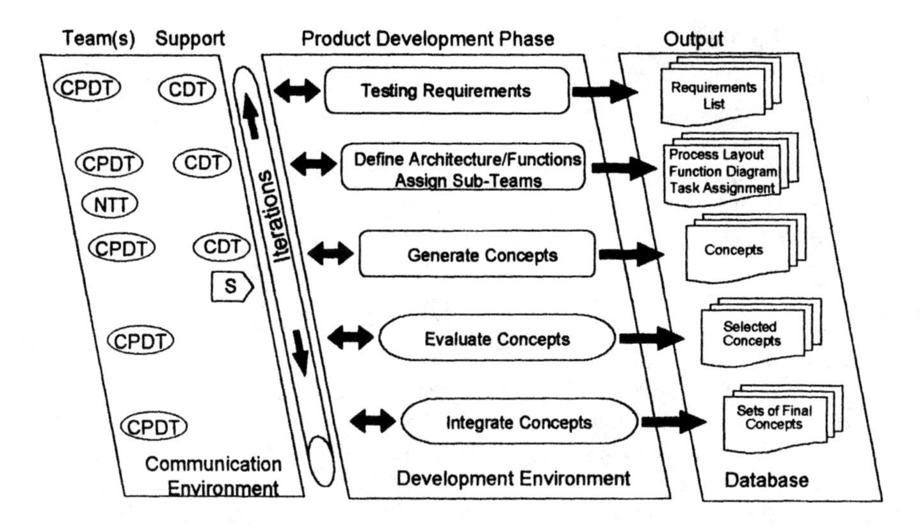

Figure 5.7 The conceptual design phase of the testing development model [Adapted with permission from J. Phair].

Table 5.9 Output for the design step: testing requirements.

Title	Requirements list
Definers: Primary	Cross-functional process development team
Definers: Support	Cross-functional product development team
Overview	Requirements list and target values Measurements needed to ensure product quality and functionality Accuracy required of all measurements Environment that is needed for testing product Data requirements for product test Identification of faulty products
Documentation	Rationale for requirements and associated target values

5.2.3 The Packaging Model

The model for the development of the packaging methods and/or line is shown in Figure 5.8. This model differs from the manufacturing process model in that there is one additional design step, packaging requirements, which is the first design step. The line layout design step may or may not be needed based on whether the packaging will be part of the manufacturing process line, or will be a standalone line. Also notice that it is assumed there is only one team working on these design steps, which are members of the CPDT. All the steps and outputs for this model are identical to those in the manufacturing process model. Therefore, we will only discuss the step that is added, which is packaging requirements.

In the packaging requirements step, the team needs to determine how the product will be packaged, stored, and shipped to the customer. There are many issues that need to be addressed in this design step. First, what type of environment is this product going into: retail, wholesale, or to another company for inclusion in one of their products? How is the product to be shipped: truck, air, train, or ship? Are there any special environmental conditions that the product requires? Is the product to be shipped to different sites, having different packaging requirements? How will the packaging material be disposed of: is it recyclable, reclaimable? These and other questions need to be addressed as shown in the output for this step in Table 5.10.

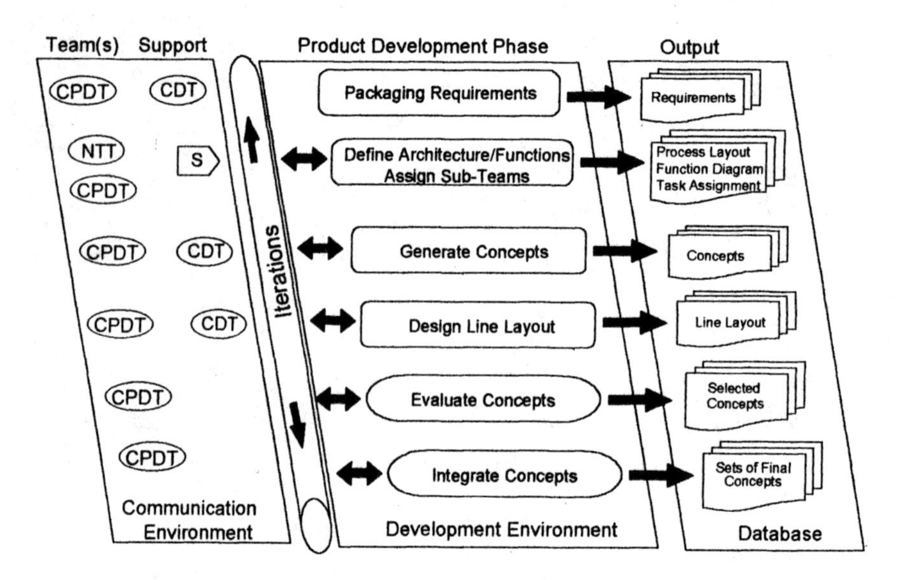

Figure 5.8 The conceptual design phase of the packaging development model [Adapted with permission from J. Phair].

Table 5.10 Output for the design step: packaging requirements.

Title	Requirements list
Definers: Primary	Cross-functional process development team
Definers: Support	Cross-functional product development team
Overview	Requirements list and target values Type of packaging required: consumer, wholesale or protective for inclusion in another product Environmental requirements of product shipment Type of shipment used: air, rail, truck, or ship Environment of product in shipment Must withstand drop testing Do end users have different requirements: language, color, materials? Disposal, reuse, or reclamation requirements of packaging material
Documentation	Rationale for requirements and associated target values

Now we have seen the conceptual design phase for all three models: manufacturing process, testing, and packaging. As discussed in the last chapter, the conceptual design phase is one of the most time-consuming phases; however, this time is well spent, as changes later in the process are minimized, leading to reduced development time and costs. The CPDT is encouraged to use concepts and solutions from previous projects, which can also reduce time and effort. These benefits can only be realized, however, if all team members generate the output required at the end of each design step, thereby recording the history and decisions of previous projects.

The final step of the conceptual design phase is the production of the milestone documents, which need to be assembled for management team approval. The major purpose of this milestone is to ensure that the best concepts have been selected for manufacturing, testing and packaging the new product, that they integrate well with each other, and that they should be able to efficiently produce the new product and have it ready for shipment. The output from the integrate concepts design step should be used for this milestone document accompanied by a description of the concepts. In the next phase, these concepts will take on a physical form.

5.3 THE DESIGN PHASE

In the design phase, the CPDT generates the final design of the manufacturing line, testing equipment and stations, and the packaging equipment and/or line. In this phase, the integrated concepts generated in the last phase will be designed in detail and in some cases with the use of prototypes will take on a physical form for testing. The final design of the line(s) should contain all information necessary for the acquisition or fabrication and installation of all equipment. As in the last section, there are three models: one for the manufacturing process, one for testing and one for packaging. The model for the manufacturing process and packaging are the same, and shown together in Figure 5.9. There are nine design steps in this phase. The testing model differs in that it has one less step, line simulation. It is assumed that testing equipment and/or stations will be integrated into the manufacturing process line, and will not be a separate line. Therefore, the line simulation design step is not needed in the testing model. This model is shown in Figure 5.10. Since these models differ little, the steps for the models will be described together. However, in the next chapter, tailoring of the models will be handled separately, as we discussed in the section on conceptual design. The tailoring process is handled separately as the requirements for each of the models could be different. Next, a description of each of the steps and their outputs for the model shown in Figure 5.9 will be given.

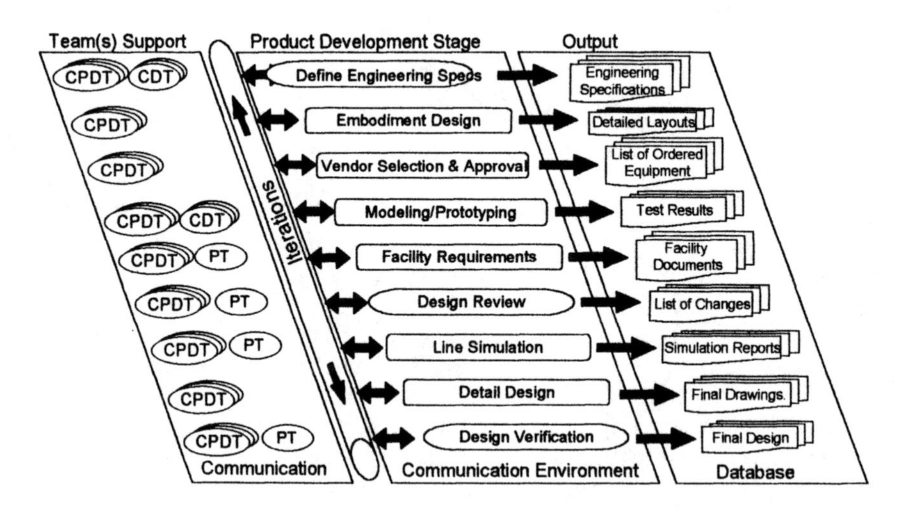

Figure 5.9 The design phase of the manufacturing process development model and the packaging development model [Adapted with permission from J. Phair].

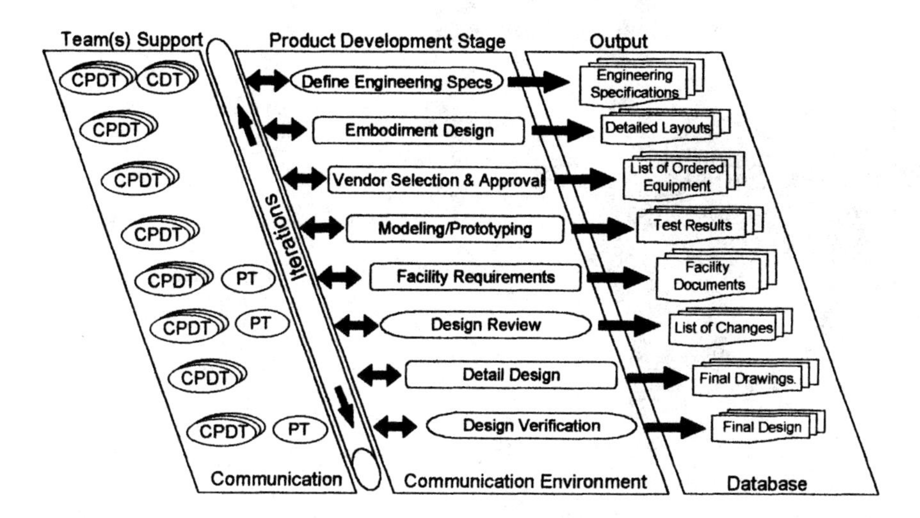

Figure 5.10 The design phase of the testing development model [Adapted with permission from J. Phair].

5.3.1 Define Engineering Specifications

In this design step, the process specifications developed in the last phase will be used as a starting point for defining the engineering specifications. This oval shape indicates that communication is key, and the CPDT should ensure that it communicates with the other teams involved in the design, especially the CDT. Any process specifications without associated values must now be defined to include target values. Furthermore, since we now have selected concepts for the design, specifications pertinent to those concepts should be developed with their associated target values, including criteria for materials, finishes, and tolerances. Tables 5.11, 5.12, and 5.13 show the outputs for the manufacturing process model, the packaging model, and the testing model, respectively. The checklist in the overview section should help the team define the engineering specifications more completely.

5.3.2 Embodiment Design

Embodiment design fills out the details of the chosen concepts and is what most engineers think of as real design. In this book, we will only briefly describe this design step. Several books cover this step in detail and are listed at the end of this

Table 5.11 Output for the manufacturing process design model for the design step: define engineering specifications.

Title	Engineering specifications
Definers: Primary	Cross-functional process development team
Definers: Support	Cross-functional product development team
Overview: Specifications	
Size Requirements	Line layout
	Station footprints
	Equipment within each station
	Area between stations
Line flow requirements	Flow of product between stations
	Motion of product
	Positioning of product in each station
Material requirements	Strength
	Impact resistance
	Corrosion resistance
	Heat resistance
	Deformation
	Expansion
	Durability
	Life
	Cost
	Recyclability
Safety	Station operation
	Materials used in stations
	Line environment
Ergonomics	User interface for product manufacture
	Repetitive motion
Assembly	Time
	Skills
	Equipment
Transport	Vibration resistance as product moves through line
	Environment control on line: heat, humidity, dust
Operation	Conditions under which line operates
Re-work	Disassembly, remanufacture, recycling
Maintenance	Replacement parts
	Service
Documentation	Rationale for target values

Table 5.12 Output for the packaging design model for the design step: define engineering specifications.

Title	Engineering specifications
Definers: Primary	Cross-functional process development team
Definers: Support	Cross-functional product development team
Overview: Specifications	
Size Requirements	Line layout
	Station footprints
	Equipment within each station
	Area between stations
Line flow requirements	Flow of product between stations
	Motion of product
	Positioning of product in each station
Material requirements	Strength
	Impact resistance
	Water and weather resistance
	Heat resistance
	Deformation
	Durability
	Life
	Reusability
	Cost
	Recyclability
Safety	Station operation
	Materials used in stations
	Line environment
Ergonomics	User interface for product manufacture
	Repetitive motion
Assembly	Time
	Skills
	Equipment
Transport	Vibration resistance as product moves through line
	Environment control on line: heat, humidity, dust
Operation	Conditions under which line operates
Maintenance	Replacement parts
	Service
Documentation	Rationale for target values

Table 5.13 Output for the testing design model for the design step: define engineering specifications.

Title	Engineering specifications
Definers: Primary	Cross-functional process development team
Definers: Support	Cross-functional product development team
Overview: Specifications	
Station Requirements	Product size
	Station footprints
	Equipment within each station
	Area between stations
	Special requirements such as vibration isolation
Line flow requirements	Test cycle time
	Motion of product
	Positioning of product in each station
Test requirements	Product measurements
	Product functionality measurements
	Measurement accuracy
	Measurement precision
Safety	Station operation
	Materials used in stations
	Line environment
Ergonomics	User interface for product test
	User interface for results interpretation
	Repetitive motion
Calibration	Time
	Skills
	Equipment
	Frequency
	Precision
	Accuracy
	Check of calibration success
Transport	Vibration resistance as product moves through line
	Environment control on line: heat, humidity, dust
Operation	Conditions under which line operates
	Rejection of failed product
Maintenance	Replacement parts
	Service
Documentation	Rationale for target values

chapter. Embodiment design depends on the design type undertaken, however, a general process was shown in Figure 4.13 that can be followed for this design step. Communication is key to this design step, and the team members should stay up-to-date with decisions among themselves and communicate regularly with the other sub-teams. As shown in Figure 4.13, the designers should first start with the main functions of the design, then move to the sub-functions. For these functions they first should consider using off-the-shelf components, and then if none can be found to fulfill the requirements, then they should design that component. In either case, at the end of this process the necessary equipment and materials for the fabrication and installation of the manufacturing stations must be selected. There are several issues that need to be considered in this design step:

- What is available from suppliers?
- What equipment can be purchased off-the-shelf rather than designed?
- Should we out-source the design of the machine to a vendor that has experience designing a particular type of manufacturing station?
- Should we out-source the fabrication of a design that must be designed in-house, or do we fabricate on-site?
- How is the equipment or parts going to be tested to ensure they meet specifications?
- What are the acceptance criteria for machines or parts purchased or designed by vendors?
- Are there any special considerations that have to be taken into account such as special transportation for oversized machines or sensitive parts? If the machine is very large, can we get it to the manufacturing line? How will the part or equipment be moved into place; will it require riggers?

The output from this step should be a final layout with a bill of materials. Table 5.14 shows the output requirements for this step.

5.3.3 Vendor Selection and Approval

In this design step, the design team works with procurement personnel to find appropriate vendors to supply those materials and equipment selected in the last design step. This is a key step in the design process, and there are many criteria to be considered in the selection. These criteria include the ability to provide needed equipment on schedule, with required quality, lead times, and cost. In addition, procurement must determine how the machines or equipment will be purchased, what terms of payment and delivery can be negotiated, what the terms of equipment acceptance will be, and what services the vendor will provide in the delivery, set-up, and maintenance of the equipment. These key decisions should be documented in the output, as shown in Table 5.15. This is a particularly important step to document for future projects, as the work done in this step can save time in ensuing projects.

Table 5.14 Output for the design step: embodiment design.

Title	Detailed layouts
Definers: Primary	Cross-functional process development team
Definers: Support	
Overview	Detailed layouts
Documentation	Preliminary layouts • General arrangement • Components • Shapes • Materials
	Detailed layouts • Major components and relationships • Components • Calculations • Standards and regulations

Table 5.15 Output for the design step: vendor selection and approval.

Title	List of ordered equipment
Definers: Primary	Cross-functional process development team
Definers: Support	
Overview	Station or equipment name
	Chosen supplier
	Equipment to be delivered
	Cost
Documentation	Equipment layout drawings
	Lead time for equipment
	Rationale for choosing vendor
	Supplier information: name, address, phone number and contact names
	Terms of agreement including what equipment is to be delivered, delivery method, set-up and testing method, repair, and maintenance agreements
	Approved alternate suppliers Part cost information Contact information Lead time
	Suppliers of parts that were not approved

5.3.4 Modeling and Prototyping

This design step is performed selectively on those processes that are new and/or critical to the manufacture, test and packaging of the new product. The designers may choose to either model some aspect of the design as a way in which to virtually test it, or they may decide to build a prototype. These prototypes are used to test the functionality of a particular aspect of a station and to ensure that the resulting part is acceptable from a functional and quality aspect. In general, the models or prototypes will only test one aspect of a station, and will not be a fully functional prototype. As discussed in Chapter 4, the three main purposes of these models or prototypes are to enhance the understanding of the design team, to communicate the functionality to other team members and to the CDT, and to indicate a milestone in the design. Only those functions that are considered critical should be modeled or prototyped and tested. Modeling or prototyping every function of a station would be too time consuming and lead to scheduling delays. The team must carefully consider what functions need testing and what tests will be performed to ensure satisfactory operation. Table 5.14 shows the output for this design step.

Table 5.16 Output for the design step: modeling/prototyping.

Title	Test results
Definers: Primary	Cross-functional process development team
Definers: Support	Cross-functional product development team
Overview	Brief description of the models or prototypes and findings of tests performed
Documentation	Rationale for each model or prototype developed Information desired Test plans
	Models and/or prototypes
	Results of tests

5.3.5 Facility Requirements

In this design step, the CDPT must decide what modifications or additions to the facility housing the manufacturing line must be made to accommodate the equipment. Using the outputs from the design steps, integrate concepts, embodiment design, and vendor selection and approval, the design team must decide where the electrical, air, and water access points should be placed in addition to ventilation requirements. The designers must also consider any special requirements that the equipment may require including vibration isolation, floor support, and lighting.

This step is critical to the design of the line, and for ease of installation. The output for this step is a drawing that will be given to the facilities group for implementation. This key step should be performed in a timely manner to ensure that the facility is ready for equipment installation. Table 5.15 shows the output for this step.

Table 5.15 Output for the design step: facility requirements.

Title	Facility documents
Definers: Primary	Cross-functional process development team
Definers: Support	Product team
Overview	Facility drawings and documents
Documentation	Rationale for changes to existing facility
	Drawing containing facility requirements such as power, light, air, ventilation, etc.

5.3.6 Design Review

This step is one of the most critical in this phase. In this step, the CPDT and PT should have a formal meeting in which all designed components, stations, layouts, and prototyping results should be reviewed. The main purpose of this review is to ensure that the design can be integrated into an operational line that can produce, test, and package a quality product. The designs should be checked to ensure that they are complete with no missing information. Any integration problems found should be noted and fixed. This step often precipitates a return to embodiment design. This is a key step in the design process, and should not be overlooked. This is the time to make any final changes to the design to ensure that it meets product and engineering requirements, and that the product team is happy with the design, as they will be responsible for the support of the line once it is operational. The main output of this step are changes to be made to accommodate the manufacture of the product and integration of the line. Table 5.18 shows the output for this step.

5.3.7 Line Simulation

Now that the design of the manufacturing line has been integrated, reviewed, and the changes approved, a computer simulation of the operation of the line can be performed. This step may not be needed for the packaging design model, if it is to be a part of the manufacturing process line, and not a stand-alone line. The test stations should be part of the manufacturing process line simulation, as these sta-

Table 5.18 Output for the design step: design review.

Title	List of changes
Definers: Primary	Cross-functional process development team
Definers: Support	Product team
Overview	List of changes to be made to meet specifications and to accommodate station integration, manufacture of product, and line support
Documentation	Evaluation results and rationale behind suggested changes
	Parts and documents affected by changes

tions should be integrated into the manufacturing lines. Not all companies have the software to perform this step; however, it can be worth the money and effort to find an appropriate package for the company. Some of the less expensive personal computer packages can fulfill the needs of a small company better than the more expensive UNIX-based packages. The main purpose of this design step is to test the line balance, and to identify bottlenecks that can significantly slow down production. The line simulation can also be used to identify maintenance strategies and estimate the number of personnel needed to service the line. The product team is in a support role in this step, as they can provide valuable input for maintenance data. In addition, they will be responsible for the line once it is producing product, therefore, they have a vested interest in seeing a well-balanced, functional line. Table 5.19 shows the output of this design step.

Table 5.19 Output for the design step: line simulation.

Title	Simulation reports
Definers: Primary	Cross-functional process development team
Definers: Support	Product team
Overview	Brief description of the simulation and findings of tests performed
Documentation	Rationale for input data for stations and line including cycle times, estimated down time, maintenance, station mean time between failure times

5.3.8 Detail Design

Now that the design has been reviewed, and any problems that were detected have been rectified, the final form of the design can be completed. In this design step, all detail and assembly drawings should be completed along with the bill of materials. In the detail drawings, the dimensions should be toleranced, and part materials and manufacturing instructions should be completed. In the assembly drawings, the drawing should show how the parts fit together, and each part should have a number that corresponds with an entry on the bill of materials. These drawings should have enough detail that the equipment can be fabricated, installed, set-up, and run. In general, most companies have their own standards for drawings. These standards should be outlined here for the team members to follow. Table 5.20 shows the output for this design step.

Table 5.20 Output for the design step: detail design.

Title	Production documents
Definers: Primary	Cross-functional process development team
Definers: Support	
Overview	Detail drawings
	Assembly drawings
	Bill of materials
Documentation	

5.3.9 Design Verification

The goal of this design step is to check all drawings and documentation for completeness and for sign-off on the drawings by the appropriate parties. This step should be accomplished through a series of meetings with the appropriate signatories and those responsible for the drawings. Table 5.21 shows the output for this design step. If the drawings must be changed before acceptance, these changes should be noted and justified in the documentation. This is the last step of the design process before production and is therefore key to finalizing the design.

At the end of this design stage is the process approval milestone, where the management team must approve the manufacturing line design. A document should be prepared that includes the engineering specifications, the results of the modeling, prototyping and simulations, and the signed and verified production documents. This document should be easily and quickly put together if the team has developed all the outputs along the process. The purpose of this milestone is to ensure that the manufacturing process design is complete and that it is ready to be handed over to the product team.

Table 5.21 Output for the design step: design verification.

Title	Final design
Definers: Primary	Cross-functional process development team
Definers: Support	Product team
Overview	Detail drawings with approval signatures
	Assembly drawings with approval signatures
	Bill of materials with approval signatures
Documentation	Documentation for any changes needed to the drawings before approval can be made
	Rationale for changes to drawings

5.4 THE PRODUCTION PREPARATION PHASE

This phase includes the procurement, installation, and testing of the manufacturing line, and is used to ensure that the a quality product can be produced, tested and packaged in this facility. As in the design phase, the models are the same for manufacturing process, testing, and packaging. Therefore, all the steps and their outputs from the three models will be discussed together rather than separately. This phase consists of ten design steps: procurement, installation, equipment integration, equipment acceptance, debug, documentation, training, pilot production, production validation, final testing, and release to production. This phase is managed by the cross-functional process development team with support from the product team. The phase ends with the production approval milestone. Figure 5.8 shows the production preparation phase and its design steps and outputs.

5.4.1 Procurement

The procurement design step is the culmination of earlier work in which long, lead-time parts are brought in to the facility from the vendor. Some of these parts may have been ordered in the vendor selection step in the design phase. In addition to these long, lead-time parts, made-to-order equipment that has been chosen as part of the station designs is also brought in. Finally, off-the-shelf parts which may have been selected in the design phase are acquired. This step should be fairly straightforward since all the information required in this design step should have been considered in earlier design steps. This design step is used to ensure that all equipment has been considered, and that all supplier information is up to date. Table 5.22 shows the output from this step.

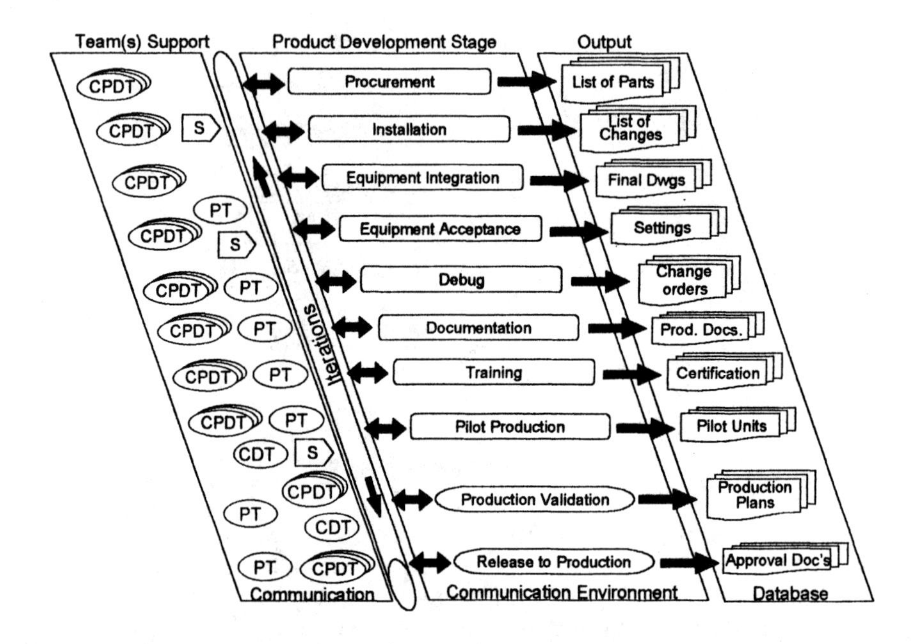

Figure 5.11 The production preparation phase for the manufacturing process model, the testing model, and the packaging model [Adapted with permission from J. Phair].

Table 5.22 Output for the production preparation step: procurement.

Title	List of parts
Definers: Primary	Cross-functional process development team
Definers: Support	
Overview	Part name
	Supplier
	Part cost
Documentation	Assembly drawings containing parts
	Part Number
	Lead time for each part
	Rationale for choosing each vendor
	Supplier information: name, address, phone number, and contact names
	Approved alternate suppliers Part cost information Contact information Lead time
	Suppliers of parts that were not approved

5.4.2 Line Installation

Installation of equipment on the line can occur as soon as the equipment arrives at the company. Any problems requiring changes to the equipment should take place as soon as possible, in addition to any realignment or relocation of the equipment that is needed. The output for this step is shown in Table 5.23.

Table 5.23 Output for the production preparation step: line installation.

Title	List of changes
Definers: Primary	Cross-functional process development team
Definers: Support	Supplier
Overview	Equipment or part name
	Supplier
	List of changes required
	Update of drawings for equipment
Documentation	Rationale for equipment changes

5.4.3 Equipment Integration

Integration into the facility and between stations should take place in this step. Any changes to the line that are needed require consultation with the team members, and a list of these changes should be documented so that the design of the line matches what is on the manufacturing floor. It is easy to ignore these documentation changes, but they are crucial for the product team, which has to support the line after it has been released to production. Table 5.24 shows the output for this step.

Table 5.24 Output for the production preparation step: equipment integration.

Title	Final drawings
Definers: Primary	Cross-functional process development team
Definers: Support	
Overview	Final drawings of manufacturing line
Documentation	Rationale for line changes

5.4.4 Equipment Acceptance

In this production preparation step, the equipment that has been installed is tested to ensure that it is operating as designed. The acceptance criteria should have already been documented in the design phase either in engineering specifications step or in the vendor selection step. The acceptance of all materials and machines requires signatures of the responsible parties both in the CPDT and in the PT, who are the personnel who will be supporting the equipment when it is operational. In addition to the approval signatures, the settings on the machines for running the test parts should be documented. The output for this step is shown in Table 5.25.

Table 5.25 Output for the production preparation step: equipment acceptance.

Title	Settings
Definers: Primary	Cross-functional process development team
Definers: Support	Product team, supplier
Overview	Acceptance documents and signatures
	Equipment settings
Documentation	Documentation of tests run and outcomes

5.4.5 Debug

This is one of the most important steps in the production preparation phase, and can be one of the most time consuming. Often management assumes that once the equipment is installed and integrated that the line should be operational. This is especially true of a line design that may be identical to one already in operation in the factory. However, once equipment is moved, it must be run and debugged to iron out the problems to produce a quality product. This step can begin once all the equipment is in place and accepted. Often the same few products will be run down the line several times, requiring that the team members disassemble the products if possible and reuse the parts, as the product produced at this stage cannot be shipped to customers and is a write-off financially. Any problems found at this stage need to be documented and fixed, and the changes documented in the final drawings. The product team should support the CPDT in this step. The output is shown in Table 5.26.

5.4.6 Documentation

In this step, documentation, which should have been started in earlier steps such as equipment acceptance and debug, is put together and includes such items as the

Table 5.26 Output for the line production preparation step: debug.

Title	Change orders
Definers: Primary	Cross-functional process development team
Definers: Support	Product team
Overview	Change orders for any design changes needed
Documentation	Documentation of test runs for line debug
	Documentation of any problems noted in the line
	Rationale for changes in the equipment
	Updated drawings for the equipment and line

equipment operation requirements, maintenance, and machine settings. In addition, the team should ensure that any information required to operate the equipment to produce a quality product should be noted. The product team needs to be actively involved with the CPDT in this process. This is a very important step in that it can also serve as the documentation needed for ISO certification of the line. The output for the three models are shown in Tables 5.27, 5.28, and 5.29.

5.4.7 Training

Now that the line is operational and the documentation complete, training can begin on the line. The personnel who will be responsible for operating and maintaining the line are trained and certified when training is complete and the team finds them competent to operate the line. The workers should be familiar with the set-up of the equipment, changeover operations, minor maintenance, and the processes for which they are responsible. They should also understand when maintenance needs to be called in to maintain or repair machines. Finally, they are ultimately responsible for the quality of the product coming off the line. Therefore, they need to be able to recognize quality problems and how to fix them. The line workers are responsible for upholding quality and ISO standards. The output for this step is shown in Table 5.30.

5.4.8 Pilot Production

Once the product and the processes are ready, pilot production runs can be performed. This step is carried out with all teams in place to support and document the process and any problems that may occur. These teams include the CPDT, the CDT, the PT, and support from the supplier. This step is identical to the step performed in the product development model. Therefore, these teams need to coordinate the timing of the pilot production runs. Pilot production may not occur on all products; it depends on the volume of products produced. However, if the volume is high (>100), then it is recommended that the company produce products

Table 5.27 Output for the manufacturing process model's production preparation step: documentation.

Title	Production documents
Definers: Primary	Cross-functional process development team
Definers: Support	Product team
Overview	Quality policy Objectives Commitment Customer expectations
	Responsibility and authority
	Resources: personnel
	Quality system procedures
	Documentation control plans
	Internal audit plans
	Maintenance plan for documents
	Review of documents
Documentation	Procedural documents Product identification and traceability Procedural documents for each station Compliance with standards, codes, and quality plan Monitoring and control of process parameters and production characteristics Criteria for work done in the stations Standards, samples, and/or illustrations Maintenance plan for each station and the lines

Table 5.26 Output for the testing model's production preparation step: documentation.

Title	Production documents
Definers: Primary	Cross-functional process development team
Definers: Support	Product team
Overview	Quality policy Objectives Commitment Customer expectations
	Responsibility and authority
	Resources: personnel
	Quality system pProcedures
	Documentation control plans
	Internal audit plans
	Maintenance plan for documents
	Review of documents
Documentation	Procedural documents Test procedure for product Identification of faulty product Calibration procedure Control of identified faulty product Maintenance of equipment Equipment safe guard
	Corrective action Correction or product problems Review of product retest

Table 5.29 Output for the packaging model's production preparation step: documentation.

Title	Production documents
Definers: Primary	Cross-functional process development team
Definers: Support	Product team
Overview	Quality policy Objectives Commitment Customer expectations
	Responsibility and authority
	Resources: personnel
	Quality system procedures
	Documentation control plans
	Internal audit plans
	Maintenance plan for documents
	Review of documents
Documentation	Procedural documents Handling to prevent damage Storage procedures Delivery Procedural documents for each station Compliance with standards, codes and quality plan Monitoring and control of process parameters and production characteristics Criteria for work done in the stations Standards, samples, and/or illustrations Maintenance plan for each station and the lines
	Corrective action Correction of packaging problems

Table 5.30 Output for the production preparation step: training.

Title	Certification
Definers: Primary	Cross-functional process development team
Definers: Support	Product team
Overview	Certification for trained personnel
Documentation	Training manuals
	List of any problems encountered or changes to procedures or documents

on the full production line. In these runs, all stations should be up and operational. This is the time to test how the line performs in the manufacture of the new product. These runs are used to find any remaining problems with hardware, software, or parts, and to make any necessary adjustments to the stations and the line to ensure that it works smoothly. The product made on these runs should be checked for functionality, quality, and compliance with all requirements. These products cannot be shipped to customers, but the company may allow them to be used internally. The output for this step is shown in Table 5.31.

Table 5.31 Output for the production preparation step: pilot production.

Title	Pilot units and fesults
Definers: Primary	Cross-functional process development team
Definers: Support	Cross-functional product development team, Product team, supplier
Overview	Resulting pilot products
	List of problems encountered and solutions proposed and undertaken
	Production fate
Documentation	List of problems Proposed solution Personnel responsible Schedule for solution implementation Date of completion
	Product quality check Results Recommendations Schedule for implementation if needed Responsible person

5.4.9 Production Validation

This is the final test of the manufacturing line before full scale production occurs, which is a phase that will be discussed at the end of this chapter. In this step, the product development team takes the lead, and the CDT and CPDT act as support teams. This is the start of the hand-off of line responsibility from development to production. In this design step, the product that comes off the line is checked by a product assurance group to ensure that it meets all design requirements. This group should check tolerances, functionality, and quality. The product team needs to set the frequency of checks to be done on the line by line personnel to ensure

Table 5.32 Output for the production preparation step: production validation.

Title	Production plans
Definers: Primary	Product team
Definers: Support	Cross-functional product development team, cross-functional process development team
Overview	Report of production run results
Documentation	Separate production plan for each production area Requirements for that section of production Characteristics of the product produced in that area Tolerances to be checked Frequency of checks Inputs and output to the section Referenced documents
	Safety documents for each section Material safety data sheets Emergency plan and documentation

that the processes stay within process ranges. All documentation should be checked for completeness, especially safety documents. At this stage of the production preparation phase, all work should be complete, and the line should be ready for production. The output for this step is shown in Table 5.32.

5.4.10 Release to Production

This final step of the production preparation phase should be more symbolic than a working one. This is the formal hand-off of the production line to the product team, who is now responsible for production. They must formally sign-off on all documents, and make any last request for changes, which should be none since all these problems should have been fixed in the pilot production and production validation runs. Table 5.33 shows the output for this step.

At the end of this design step, the production approval milestone takes place. At this milestone, the management team needs to guarantee that all parts are ready and available for production, that all changes needed to the product and the production line are completed, and that all production documents are available. Again, as in the previous milestone, this documentation should be easily gathered from the outputs of all design steps. This step should be a formality if the concurrent engineering process and teamwork have been followed. This milestone concludes the production preparation phase.

Table 5.33 Output for the production preparation step: release to production.

Title	Approval documents
Definers: Primary	Product team
Definers: Support	Cross-functional process development team
Overview	Acceptance documents and signatures
Documentation	Any changes or exceptions noted

5.5 PRODUCTION AND SERVICE PHASE

The last phase is common to the product development model, the manufacturing process development model, the testing development model, and the packaging development model. This phase consists of four steps: production, product distribution, customer support, and product retirement. This phase, shown in Figure 5.12, is mainly the domain of the product team, with support from suppliers and field service representatives. A description of these steps and their output follows.

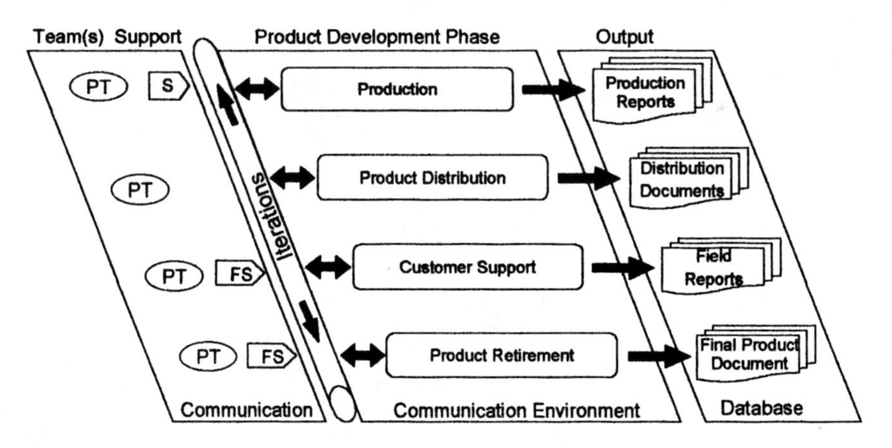

Figure 5.12 The production and service phase [Adapted with permission from J. Phair].

5.5.1 Production

In this step, the product team, which is a team that has been selected from product development and process development team members, is responsible for supporting the production lines. They assist the production personnel in solving problems, answering questions, and line maintenance. For example, suppose manufacturing experiences a problem with parts that will not fit into the feed mechanisms in some of the manufacturing process stations. It is the responsibility of the product team to assess the problem and decide if it is a parts problem (such as parts not being within specifications) or a feeder problem. After diagnosing the problem, they must then take action to correct it.

The manufacturing personnel must handle the day-to-day production issues on the manufacturing line(s). They are responsible for the production of a quality product that has been tested and packaged for shipment to the customer. As a result, they are responsible for reporting the production for the day, week, and month. They are also responsible for reporting on and fixing product quality problems and calling in the product team when needed. The output for this step is shown in Table 5.34.

Table 5.34 Output for the production and service step: production.

Title	Production reports
Definers: Primary	Product team and manufacturing personnel
Definers: Support	Suppliers
Overview	Reports broken into production sections and possibly by station Number of product produced Number of rejects Number of reworked products Scrap percentage
	Engineering change orders for lines or product as needed
Documentation	Production goals for each section Work in process created and used Goal for scrap percentages Goal for rejected and reworked product Rationale behind missed goals
	Engineering change orders Reported problem with line or product Suggested fix with rationale Changed drawings with signatures

5.5.2 Distribution

The PT also supports the activities of product distribution, fielding problems with packaging, storage, and distribution. The shipping and receiving department is primarily responsible for coordinating the distribution of finished goods. The output for this step is shown in Table 5.33.

Table 5.33 Output for the production and service step: distribution.

Title	Distribution documents
Definers: Primary	Product team and shipping and receiving personnel
Definers: Support	
Overview	Type of product shipped Quantity shipped Name of carrier Mode of shipment (truck, rail, plane, ship) Estimated date of arrival Percentage of on-time orders
Documentation	Shipment schedule Exceptions report for missed shipments Personnel responsible

5.5.3 Customer Support

This step can include the installation of a product at a customer site, support of the product during customer use, and handling reported product problems. Product complaints and a request for advice have different impacts on the company, so each is tracked in a report. These reports help the company track complaints, their severity, and if there is a problem with production of the product. If the same complaint occurs a given number of times, then a flag is raised and action must be taken to resolve this problem, which could indicate defective parts, problems on the manufacturing lines, in test, or with shipment. Furthermore, these customer interactions can give the company valuable information for future design projects, since they can point out weaknesses in the product, its manufacture, or its use. The output for this step is shown in Table 5.36.

5.5.4 Product Retirement

This step can be used to address two different issues: what the customer does with the product after its useful life is over and when the company decides to stop production of the product. Once the product is retired, the company must decide how long to support the product through field service and providing replacement parts. Table 5.37 shows the output for this step.

Table 5.36 Output for the production and service step: customer support.

Title	Field reports
Definers: Primary	Product team and field service
Definers: Support	Customer
Overview	Name of customer requesting service Reason for request
	Problem report Complaint number Cause of problem Action taken Total number of complaints to date Total number of complaints from this customer Usage questions Question asked Answer given Total number of this question asked to date Total number of questions from this customer
Documentation	Detailed reports How will problem impact relationship with customer? Who was contacted within company to assist in solving problem? Ways to improve product, production, or documentation

Table 5.37 Output for the production and service step: product retirement.

Title	Final product document
Definers: Primary	Product team
Definers: Support	Field service
Overview	Product end-of-life Final destination of product after useful life Documentation regarding any potential hazards in product disposal, re-use, recycling or re-mfg Logistics of handling unusable product
	Production end-of-life When to stop production Rationale behind decision Product support decisions How long will product be supported What parts need to be kept on hand for support
Documentation	Production end-of-life Dates for suppliers to stop shipment of components Number of products to keep on hand for future repairs

5.6 SUMMARY

The process development model consists of three sub-models: manufacturing process development, testing development, and packaging development. In turn, these models consist of four design phases, which include project planning, conceptual design, design, and production preparation. These development phases are carried out primarily by the cross-functional process development team with support from the new technology team in the early design steps and from the cross-functional product development team with regards to product information. In addition, field service, suppliers, customers, and the product team, in the later design steps, are also involved in these phases. Each design phase follows the same general layout in which teams and support members are assigned to be responsible for various design steps. These teams then carry out the design step that results in decisions and outputs for that step which are recorded in a database. The decisions made within each phase begin very broad, but with time and iteration within the phase, the decisions become more detailed until a final milestone report for that phase is completed and ready for management approval. Each of the outputs has been detailed in this chapter, and each of the milestones has been overviewed. The final phase of the product development model and the manufacturing development model is also discussed in this chapter. This phase, production and service, is the culmination of all the other phases, the one in which the product is produced and shipped to the customer. It concludes the development of the product and its manufacture.

REFERENCES AND BIBLIOGRAPHY

Henson, C., *Defining the Outputs of a Small Business Concurrent Engineering Model*, Senior Thesis, University of Virginia, 1998.

Kemser, H., *Concurrent Engineering Applied to Product Development in Small Companies*, Masters Thesis, University of Virginia, 1997.

Phair, J., *Integrating Manufacturing and Process Design into a Concurrent Engineering Model*, Masters Thesis, University of Virginia, 1999.

SUGGESTED READING

Engineering Design by G. Pahl and W., Springer-Verlag, New York, 1996.: An in-depth text that covers product planning through embodiment design, with special concentration on a systematic approach to conceptual and embodiment design.

Total Quality Development by Don Clausing, ASME Press, New York, 1994: Covers concurrent engineering from a quality perspective.

Quality Function Deployment by Ronald Day, ASQC Quality Press, Milwaukee, 1993: A very complete book on the use of the House of Quality.

Total Design by Stuart Pugh, Addison-Wesley, Reading, MA, 1990: Covers Pugh's approach to systematic design. He is known for his development of the concept selection chart that was adapted in this book.

Total Quality Development by Don Clausing, ASME Press, New York, 1994: Covers concurrent engineering from a quality perspective.

The Mechanical Design Process by David G. Ullman, McGraw Hill, Boston, 1997: Covers the design process from a mechanical engineer's point of view.

Product Design and Development by K. Ulrich and S. Eppinger, McGraw-Hill, Inc., New York, 1995: Focuses on methods, tools and techniques for design teams to use in the development of new products.

6

Tailoring the Methodologies

In collaboration with Hans-Peter Kemser and Johanne Phair

The most unique aspect of this book is the tailoring process discussed in this chapter. Through our observation and work with small companies using concurrent engineering, we found that the team members became very frustrated when required to go through all steps for every project whether the steps were needed or not. This frustration lends itself to sloppy work and an apathetic view of concurrent engineering. Therefore, we developed a method for tailoring the methodology based on the design project characteristics. In this chapter, we will discuss the major types of design that a small company will encounter, both in product design and process design. Then, we will present models for each of these design types. Finally, we will introduce a method that will allow a team to customize the product and process design methodologies to suit their projects, thereby eliminating needless steps from the process.

6.1 TAILORING THE PRODUCT DESIGN METHODOLOGY

The tailoring process is done in the project planning phase of the product design methodology. The process is shown in Figure 6.1, and begins with the assessment of the project, which involves input from four main groups: engineering and design, manufacturing, sales and marketing, and resource management. Based on this assessment, a product development target is plotted, and then the product development model is created for this project. Finally, the model is customized based on the company's and the project's characteristics and needs. Each of these steps is discussed in detail in the next few sections.

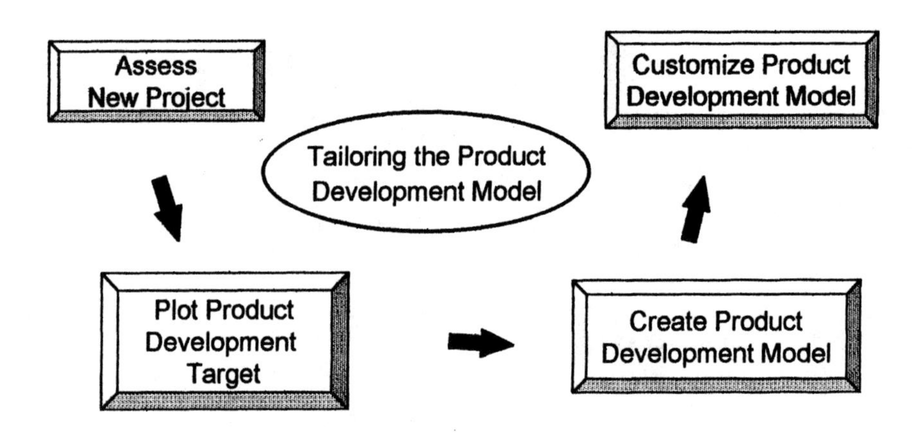

Figure 6.1 Illustration of the steps of the tailoring process [Reprinted with permission from H. Kemser].

6.1.1 New Project Assessment

The assessment of a new project is a subjective evaluation in which the cross-functional product development team (CDT) members compare the new project to those already finished or underway within the company. There are no clear-cut rules on how to rate each criterion, but rather the judgment of the team members and their knowledge of the company is used. Therefore, each company will evaluate a project differently since it will be compared with other projects within that company.

There are twelve project assessment criteria for tailoring the product development methodology, as shown in Table 6.1. These criteria are divided into four main categories: engineering, manufacturing, sales and marketing, and resource management. Each criterion is rated as level A, B, or C, with the maximum and minimum ranges for these levels on the left and right, respectively. In the far right column is a space provided for comments about the level selection.

The twelve criteria shown in Table 6.1 pertain only to the tailoring process. These criteria do not include other important product planning issues such as team formation, project priority, project cost, or risk assessment. These issues should be addressed in project planning, but are not a part of the tailoring process. Next, we will discuss each of the criteria in detail.

Engineering

Within this category, there are three criteria that must be evaluated for the tailoring process: design paradigm, product complexity, and standards and specifications.

Table 6.1 Assessment table for the tailoring process of the product development model [Reprinted with permission from H. Kemser].

Section #	New Project Assessment Criteria		Level* A	B	C		Comments:
Design							
1	Design Paradigm	Original				Incremental	
2	Product Complexity	High				Low	
3	Standards and Specifications	High				Low	
Process							
4	Product Process	New				Existing	
5	Process Complexity	High				Low	
Sales and Marketing							
6	Time-to-Market	Short				Long	
7	Sales Volume	High				Low	
Resource Management							
8	Demand for Analytical Resources	High				Low	
9	Customer Involvement	High				Low	
10	Partner Involvement	High				Low	
11	Supplier Involvement	High				Low	
12	Customer Support of Product	High				Low	

** Evaluation of the Level depends on other products that are produced within the company.*

Design Paradigm A design paradigm describes the different types of projects undertaken during the development process in a small company. Through observation and the experience of working with design teams, we found there are generally three different types of design that are undertaken during product development in small companies. These models are the backbone of the product design methodology and will be used as a basis for tailoring the methodology to fit a particular project [Ter-Minassian, 1995; Kemser, 1997]. These three design types are: original, evolutionary, and incremental.

- *Original Design*
  An original design in this book is defined as one that is new to the company and therefore has no design history associated with it. This design can be original to the market as well, but need not be. An example of an original design would be a company that has specialized in large system printers deciding to enter the personal computer printing business, and therefore developing a personal inkjet printer. Since a design history for the new product does not exist within the company, the original design model begins with the definition of functions for the product and includes all the steps in the conceptual and design phases discussed in Chapter 4. Figure 6.2 shows the conceptual and design phases for original design.

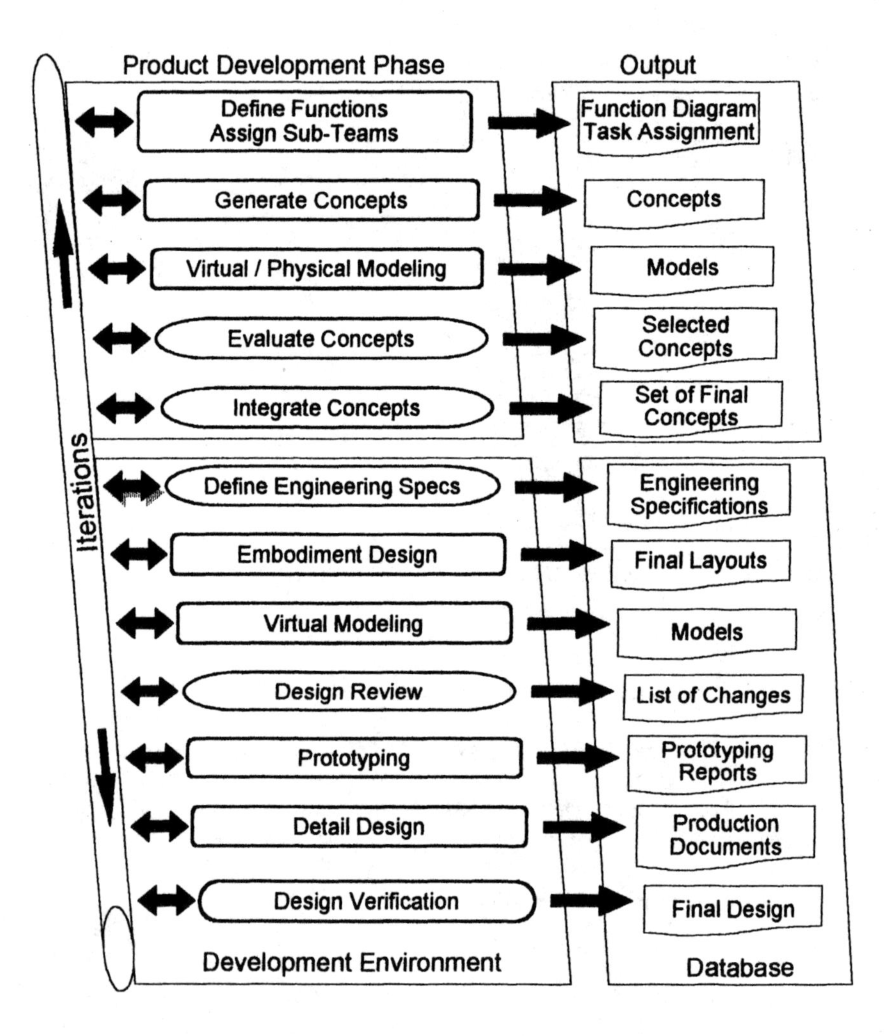

Figure 6.2 Original design model which includes the conceptual design and design phases [Reprinted with permission from H. Kemser].

This design model begins with the design step, define functions and assign sub-teams. Since a design history does not exist for this type of design project, we cannot yet define the architecture of the system and therefore begin with the definition of the functions desired in the new product. The rest of the conceptual design phase and the design phase contain all steps as defined in Chapter 4.

- *Evolutionary Design*

  An evolutionary design is the development of a new generation of products relying on earlier products as a skeleton. Therefore, design history is available to the designers, and since much of the product design already exists, many steps can be combined. An example of an evolutionary design would be a company that produces personal computer inkjet printers decides to offer a network capable printer that handles wider format paper and postscript features. Since the design is still an inkjet printer, but with additional features and functions, an evolutionary design model is used. Figure 6.3 shows the conceptual and design phases for an evolutionary design.

  The evolutionary design model begins with the design step, define architecture and assign sub-teams. Since there is a design history on previous products, then we can use the methodology described in Chapter 4 for defining architecture and assigning sub-teams. Because the skeleton of the design exists, the rest of the design steps can be combined. Although the designers will go through many of the same steps as in the previous model, there are fewer functions that need to be developed into design concepts, so the steps have been combined. In the design phase, many of the steps have been combined because less effort will be required since there are fewer new concepts that need testing. When a step is combined, then the outputs should be examined and information that is critical to the history of the design and critical decisions should still be documented. The design teams should base their outputs on those defined in Chapter 4, but they can leave out those documentation requirements that are not relevant to the design.

- *Incremental Design*

  An incremental design is the development of a new product based on an existing design. An incremental design differs from an evolutionary design in that it changes little from the previous design, only adding or removing features to fit a customer need or market niche. Because only minor or incremental changes occur in the design, and the documented architecture should already exist, the design model consists of only two steps. An example of an incremental design would be a company that produces personal computer inkjet printers decides to offer an inexpensive home model geared toward students that would still print color and black and white, but would have limited functionality at a lower price point. Since the design is still an inkjet printer, but with fewer features and functions, an incremental design model applies. Figure 6.4 shows the conceptual and design phases for an incremental design. This model has a step for generating and evaluating concepts, and one for detailing these new concepts.

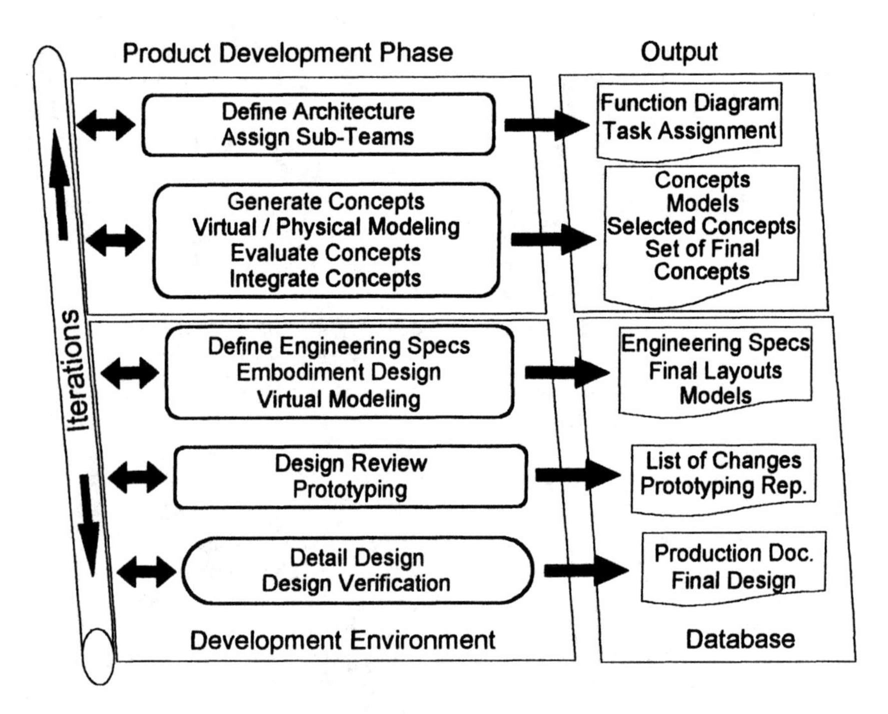

Figure 6.3 Evolutionary design model including the conceptual design and design phases [Reprinted with permission from H. Kemser].

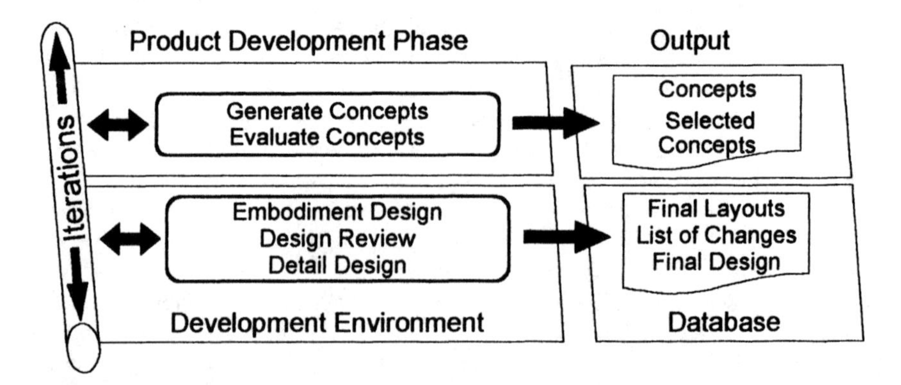

Figure 6.4 Incremental design model including the conceptual design and design phases [Reprinted with permission from H. Kemser].

Product Complexity The second criterion to be evaluated is product complexity. Although there can be many definitions for complexity, in this book we chose to consider complexity as a measure of the skill level of the personnel needed for the project. The skill level is decided by evaluating a number of issues such as: the number of components in the product, the numbers and types of technology used, the number of teams and functional areas involved, the estimated number of development hours needed, and the numbers and types of communication channels required. All these issues help decide whether the team evaluates the project as being of high, medium, or low complexity. Furthermore, complexity must be evaluated within the context of other products and projects within the company itself. What is high complexity for one company, may be low complexity for another. An example of high complexity (Level A) for a printer company would be a new printer that uses technologies not included in their other printer lines such as networking, copying, scanning, and faxing capabilities. An example of average or medium complexity (Level B) would be the introduction of some new technologies within the printer such as adding network capability to a line of inkjet printers. Low complexity (Level C) would be the development of a new printer consisting of understood technologies relying heavily on parts from other existing printer designs.

Standards and Specifications The last criterion to be evaluated in this category is standards and specifications, which addresses the type and rigidity of criteria and testing that a product must undergo. There are several types of standards that may need to be addressed such as government regulations, international standards requirements, and internal company standards. In addition, the customer or company may also impose rigid specifications that must be met and verified through testing. Again, evaluating this category is very company dependent. What one company may consider to be rigid requirements, may be typical requirements to another. If the standards and regulations imposed on the new project are significantly higher or lower than other projects within the company, then the rating would be high (Level A) or low (Level C), respectively. Again, this is a judgment call made by the team.

Manufacturing

In this category, two criteria must be evaluated for the tailoring process: production process and process complexity. The first criterion refers to whether the company has all the production processes in place or if development is needed. If a new production process is needed for the product, then it is rated as Level A; if changes to an existing process are required then it is considered Level B; if all needed processes exist with no or changes required, then it is considered Level C.

Process complexity is rated by evaluating issues such as number of production steps needed, anticipated bottle necks, number of assembled components needed, and number of parallel production activities required. An example of a product having high process complexity (Level A) would be a product that has components requiring multiple machining and surface finishing steps, since at each step there is room for quality problems and possible rework. A product having medium or average complexity (Level B) would be one with a high number of sub-assemblies and possibly some machining or surface finishing involved. A product with low process complexity (Level C) would be one with only a few components or sub-assemblies with no machining or surface finishing involved.

Sales and Marketing

There are two criteria considered in this category: time-to-market and sales volume. Time-to-market is defined as the time from the start of the product development process to the product's shipment to the customer. The team can define this time as the estimated time that the product will take to develop and deliver, or it can be defined based on when the product should enter the market according to their analysis of the situation. The first definition depends on the assigned resources and project's priority within the company. The second definition assumes that the project is a top priority, and that the resources will be assigned to the project to meet the product's ship date. The team must decide how they will define time-to-market, and then clearly state their assumptions so that project planning can accommodate these assumptions. As in other criteria, time-to-market projections of short, medium, or long are based on the time required to get other, similar products to market. Does this product have a shorter time-to-market projection compared to other similar products in the company, or is it about the same, or is it much longer? The team must make this judgment.

The second criterion, sales volume, is based on marketing's prediction of how many products will sell each year of the product's projected life. As in the previous criterion, is this volume projection higher, about the same, or lower than similar products that have been developed in the company? For example, if a company produces products that typically sell 10,000 units per year and the new product is projected to sell 50,000 units per year, then Level A (high) is selected. Again, there are no numerical guidelines for these criteria, each company must make the judgment as to what is high, average, or low for their company and for the project.

Resource Management

There are five criteria for this category: demand for analytical resources, customer input, partner involvement, supplier involvement, and customer support of the product.

The demand for analytical resources is defined as the level of analysis that is required to bring a safe, high-quality product to the market. The analytical resources would be tools such as finite element analysis, computer-aided design software, prototyping software for the product and the production lines, and simulation software for production processes. As in previous criteria, the categories are high, average, and low, and are selected based on the development of other products within the company. If more tools than usual are required for the development of the new product, then Level A (high) is chosen. On the other extreme, if few or no tools are needed, then Level C (low) is chosen. Level B (average) indicates that the tools needed are about average for the products developed in the company.

The next three criteria, customer, partner, and supplier involvement, indicate the degree to which other personnel are involved in the development process. Customer involvement should be self-explanatory. A partner is defined as a third party that joins with the company in aspects of the development of the product that the company cannot perform. For example, a company that develops cellular phones may decide to partner with a software company to deliver a phone that is e-mail and web capable. This relationship is in contrast with supplier involvement, in which a company purchases parts from the supplier, but does not enter into a joint venture with them. A high level of involvement (Level A) would indicate that the third party is part of all major decisions throughout the development process. A low level of involvement (Level C) indicates that there is no involvement by the third party or it comes very late in the process, such as in testing.

The last criterion, customer support of product, indicates what level of customer support the company will provide for the new product. For example, the company may decide to provide on-line support, product installation, product maintenance, training, repair, and/or warranty. Again, this criterion must be judged in context with the support provided for other products in the company.

The first time that a company uses this project assessment tool will be time consuming. The responsible team needs develop the guidelines for its use. These guidelines will be a starting point, and may change as more products are developed using this methodology. However, iteration is to be expected in concurrent engineering and continuous process improvement is an important part of this methodology.

6.1.1 Product Development Target

Once the project assessment has been completed, the results are plotted on a bull's-eye graph as shown in Figure 6.5. The graph is divided into the four categories discussed above: engineering, manufacturing, sales and marketing, and resource management. These categories are further divided into twelve sections that correspond to the twelve evaluation criteria. There are three rings on the graph, A, B, and C, that correspond to the levels in the evaluation table, with A being the outermost ring, and C being the innermost.

The goal of the development team is to reach the center of the target—a new product. The plotted evaluation criteria from Table 6.1 give an indication of the effort and resources required to reach the goal of a new product by their distance from the center. For example, Level A is the outside ring of the target indicating that it will take more effort than a point plotted in Level C, the innermost ring.

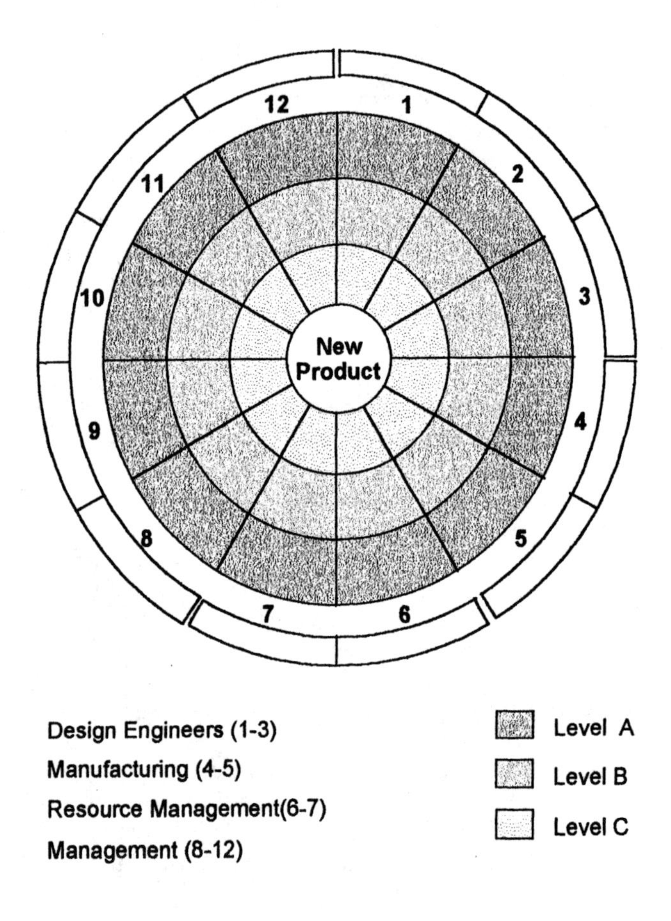

Figure 6.5 Product development target [Reprinted with permission from H. Kemser].

These plotted points will be used to create a tailored product development model for the project. The following section describes the steps to create this model, and how the criteria affect the base model.

6.1.2 Creating the Customized Product Development Model

A customized product development model is now created using the answers from the project assessment. Each criterion has an effect on the model, and will either streamline the steps or expand them depending on the needs of the project. First, the major milestones of the project are defined. Each project begins with a kick-off meeting, followed by a planning phase, which always consists of three design steps: identify needs, define product specifications, and plan development tasks. The design process ends with the production and service phase, which was discussed in the last chapter. Now we will look at how each of the criteria affect the product model.

Section 1: Design Paradigm
Level A (Original) The original design model shown in Figure 6.2 is used as the base model for this project.
Level B (Evolutionary) The evolutionary design model shown in Figure 6.3 is used as the base model for this project.
Level C (Incremental) The incremental design model shown in Figure 6.4 is used as the base model for this project.

Section 2: Product Complexity
This criterion affects the conceptual design phase. It also helps decide the number of parallel cross-functional product development teams needed and used, the degree of integration of the product needed, and the level of expertise needed within the teams.
Level A (High) Separate development steps are created for the definition of the product architecture/functions and assign sub-teams, concept integration, prototyping, and design verification. Additional cross-functional product teams are created and additional personnel with critical skills are gathered to meet the needs of the project.
Level B (Average) The concept integration and design verification development steps should be included in the model.
Level C (Low) No changes to the model are required.

Section 3: Standards and Specifications
This criterion affects conceptual design and design phases.
Level A (High) Separate development steps are created for concept evaluation, prototyping, and design verification. If the customer has had significant input into the product specifications, these stages should include the customer as a support team member.
Level B (Average) The concept integration and design verification development steps should be included in the model. If the customer has had significant input into the product specifications, these stages should include the customer as a support team member.
Level C (Low) No changes to the model are required.

Section 4: Production Process
This criterion affects the production preparation phase, and the level of integration between the product and process development teams.

Level A (New) Separate development steps are created for concept evaluation, design verification, pilot production, and production validation. Significant interaction is needed between the design teams.
Level B (Changes) The development step production validation should be included in the model.
Level C (Existing) No changes to the model are required.

Section 5: Process Complexity
This criterion affects the production preparation phase and team integration.
Level A (High) Separate development steps are created for concept evaluation, design verification, pilot production, and production validation.
Level B (Average) The production validation development step should be included in the model.
Level C (Low) No changes to the model are required.

Section 6: Time-to-Market
This criterion affects conceptual design and design phases, and the overlap and integration of the design activities. The shorter the time-to-market, then the more the tasks have to be overlapped, so as to meet the deadlines imposed. However, note that the more the tasks are overlapped the higher the risk is for cost overruns and problems.
Level A (Short) Separate development steps are created for define architecture/function and assign sub-teams, and for design verification. The steps should be overlapped as much as possible, and additional teams should be added to help speed the development process.
Level B (Average) No changes to the model are required.
Level C (Long) There is little overlap of the development steps, and there is little undertaking of risk on the project. There is little parallel development in this type of project.

Section 7: Sales Volume
This criterion affects the production preparation phase.
Level A (High) Separate development steps are created for pilot production, production validation, and field trials.
Level B (Average) The pilot production and production validation development steps are combined. If field trials are normally a part of the company's practices, then this step should be separate.
Level C (Low) No effect on the model.

Section 8: Demand for Analytical Resources
This criterion affects the conceptual design and the design phases.
Level A (High) Separate development steps are created for virtual/physical modeling in the conceptual design phase and for virtual modeling in the design phase.

Level B (Average) The virtual/physical modeling development step in conceptual design is combined with concept generation. The virtual modeling development step in the design phase is combined with embodiment design in this phase.
Level C (Low) No effect on the model.

Section 9: Customer Involvement
This criterion affects the conceptual design and the design phases.
Level A (High) The customer should be included as a support team member in the following steps: concept generation, concept evaluation, design review, prototyping, and design verification.
Level B (Average) The customer should be included as a support team member during concept evaluation and prototyping.
Level C (Low) If the customer is involved at all, it should be during prototyping.

Section 10: Partner Involvement
This criterion affects all phases of the model.
Level A (High) The customer should be included as a member of the development team.
Level B (Average) The customer should be included as a support team member for the product development team, the new technology team, and the product team.
Level C (Low) No effect on the model.

Section 11: Supplier Involvement
This criterion affects the conceptual design, design, and production preparation phases.
Level A (High) The supplier should be included as a support team member in the following steps: concept generation, concept evaluation, embodiment design, prototyping, and pilot production.
Level B (Average) The supplier should be included as a support team member during concept generation, embodiment design, and prototyping.
Level C (Low) No effect on the model.

Section 12: Customer Support
This criterion affects the production/service phase.
Level A (High) A development step is created separately for customer support.
Level B (Average) The product distribution and customer support development steps are combined.
Level C (Low) No effect on the model.

After these changes are made to the base development model, the model includes all personnel and design steps needed for a successful development project. Because the model that we have developed is general to many industries, the final step should be the customization of the model for a company's needs. In this customization, the company should not remove any steps, but adding, changing the name or the outputs is a valid customization. Removing needed design steps can lead to missing information in later steps.

6.1.3 Customizing the Product Development Model

This is the final step of the tailoring process. The purpose of this step is to streamline or expand the base model to ensure that all the needed development steps are included in the final model. Once the model is complete, no steps should be removed, as they can lead to a loss of information within the design, or may require that unnecessary iteration occur in the process, thereby slowing the product's development. The next section will show how this tailoring process is accomplished using a design example from the experience of Skalak.

6.1.4 Tailoring Process Example

A company that designs and assembles personal computers decided to offer a printer as part of their computing package. All their competitors in the PC market were large foreign companies with low assembly costs. After two years of development they offered a cost- and performance- competitive dot-matrix printer. The company did manufacture other printers, but this was their first foray into a personal computer printer. With this product's success, the company then decided to offer a wide carriage version of this printer to accommodate small businesses that need to print large forms, such as automobile parts dealers and manufacturing companies.

The company's development management team assembled a new cross-functional product team for this effort. They identified the following needs for the new product. For the first printer, the company built several new manufacturing lines; they wanted this new product to be made on these same lines. The printer needed the same look as the first printer with a comparable performance. The new printer was to be brought to the market as soon as possible, and the technology from the previous product was to be used as much as possible to shorten the development time. During the project planning phase, marketing estimated that the sales volume for the new product was 20,000 units per year for 3 years at a retail price of $499, as opposed to the first model which was estimated to have sold 100,000 units per year for 4 years at a retail price of $299. With this information, the project assessment table was completed as the first step. Table 6.2 shows the results of this evaluation.

Looking at the results of the evaluation, the design was rated as evolutionary since it was based on the previous printer, but would require many modifications. It was rated as low complexity because the technology from the existing printer would be used, and rated as average for standards and specifications since the printer was to meet UL standards and the company's own strict product standards. The product process was rated as a level B since changes were needed to allow the manufacturing lines to accommodate the larger footprint or size of the new product. The process complexity was rated as low as the product was largely assembled from supplier or in-house parts, and the in-house parts were made using injection molding, a process used extensively in the company. The new project was rated as short time-to-market, as the intent was to introduce the product as soon as possible to meet customer demand. The sales volume was rated as low compared to the other printer which had a six-fold increase in expected sales. However, limited field trials were needed to ensure customer satisfaction with the product. There were few resources needed for analysis, mainly prototyping, since this printer was largely a follow-on of the previous model. There was no partner involved in this project, and the customer was consulted at the prototype stage to ensure satisfaction with the larger footprint and the intended function of the printer. The supplier involvement was minimal, since there were modifications to only a few supplied parts to accommodate the printer's large footprint. The support of this product were to be the same as the previous product's: 1 year warranty with repair parts available for shipment, and returns accepted.

Using the results from Table 6.2, the project development target was plotted and is shown in Figure 6.6. The area within the points indicates the amount of effort required to bring this new product to market. As the figure indicates, overall the effort needed is not very high, except the demand for a short time-to-market. Next, the model for this development product will be created using the steps outlined in the last section.

First Step:

The major milestones are laid out, and the project planning phase was defined using the steps: identify needs, define product specifications, and plan develop ment tasks. Also, the production preparation phase was defined using the steps: procurement, field trials, pilot production, and production validation.

Table 6.2 Assessment table for the example.

Section #	New Project Assessment Criteria		Level* A	B	C		Comments:
Design							
1	Design Paradigm	Original		X		Incremental	Parts need modification
2	Product Complexity	High			X	Low	Use existing technology
3	Standards and Specifications	High		X		Low	
Process							
4	Product Process	New		X		Existing	Accommodate new sizes
5	Process Complexity	High			X	Low	Use existing processes
Sales and Marketing							
6	Time-to-Market	Short	X			Long	As soon as possible
7	Sales Volume	High			X	Low	~60,000 units, field trials
Resource Management							
8	Demand for Analytical Resources	High			X	Low	Few
9	Customer Involvement	High			X	Low	Prototype evaluation
10	Partner Involvement	High			X	Low	None
11	Supplier Involvement	High			X	Low	Parts modification
12	Customer Support of Product	High		X		Low	Same as other printers

** Evaluation of the Level depends on other products that are produced within the company.*

Second Step:
The first criterion indicates which basic design model was used. For this product, the evolutionary design model shown in Figure 6.3 was used.

Third Step:
The remaining criteria indicate what changes had to be made to this base model.

Section 2: No effect on the model.

Section 3: The integrate concepts and design verification development steps were included in the model.

Section 4: The production validation development step should be included in the model.

Section 5: No effect on the model.

Section 6: Separate development steps needed to be created for the definition of architecture/ functions and assign sub-teams, and design verification. The concur rent development steps were overlapped as much as the team was willing to risk. Additional development teams were considered to speed the development.

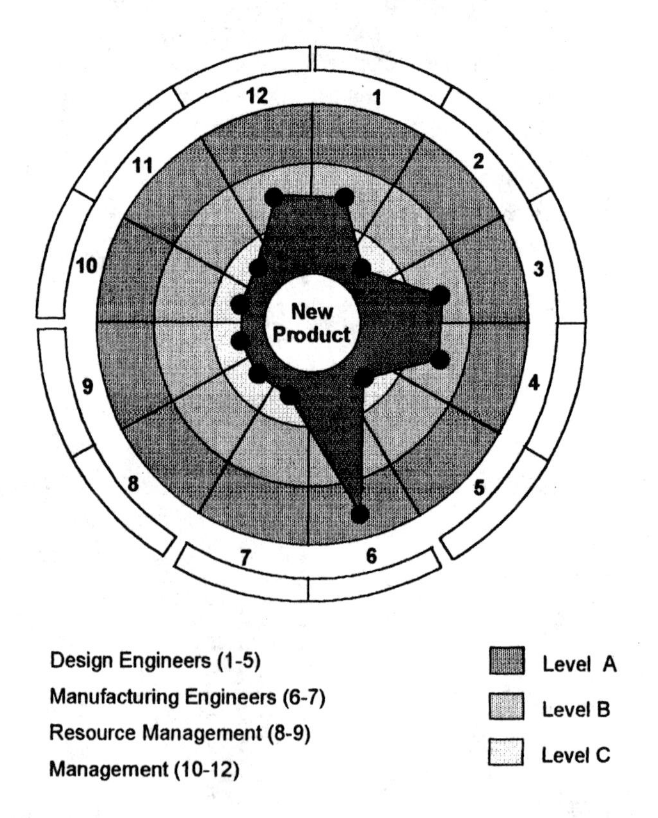

Figure 6.6 Product development target for the example [Adapted with permission from H. Kemser].

Section 7: No effect on the model.
Section 8: No effect on the model.
Section 9: The customer was a support member in the prototyping step.
Section 10: No effect on the model.
Section 11: The supplier was a support member during production validation.
Section 12: The production steps of product distribution and customer support were combined.

These modifications were made to the base model and the overview of the tailored product development model is shown in Figure 6.7. The detailed phases that were modified are shown in Figures 6.8 – 6.10.

6.2 TAILORING THE MANUFACTURING PROCESS DESIGN METHODOLOGY

In the remainder of this chapter, we will discuss how to tailor the process design methodology, which includes the manufacturing process model, the testing model, and the packaging model. The tailoring process is identical to that shown in Figure 6.1, and begins with the assessment of the project from the point of view of manufacturing, testing, and packaging. Based on these assessments, development targets are plotted for each of the manufacturing process models, and then they are customized for the new project. Each of these steps are discussed in detail in the next sections.

6.2.1 New Manufacturing Project Assessment

As in the product development assessment, the assessment of a new manufacturing project is also a subjective evaluation in which the team members compare the new project to those already finished or underway within the company. There are no clear cut rules on how to rate each criterion, but rather the assessments rely on the judgment of the team members and their knowledge of the company. Therefore, each company will evaluate a project differently since it will be compared with other projects within that company.
The project assessment tables are for process, testing, and packaging, shown in Tables 6.3-6.6. Separate tables for each of the manufacturing process development models allow the models to be tailored according to their project needs, as each might be different. As seen in the tables, there are four – categories of criteria: planning, product, sales and marketing, and resource management. As in the product development project assessment, there are three levels from which to choose, levels A, B, or C, with the maximum and minimum ranges for these levels on the left and right, respectively. The one exception is that there are four manufacturing process models from which to choose; these will be discussed below. In the far right column is a space for comments about the level selection.

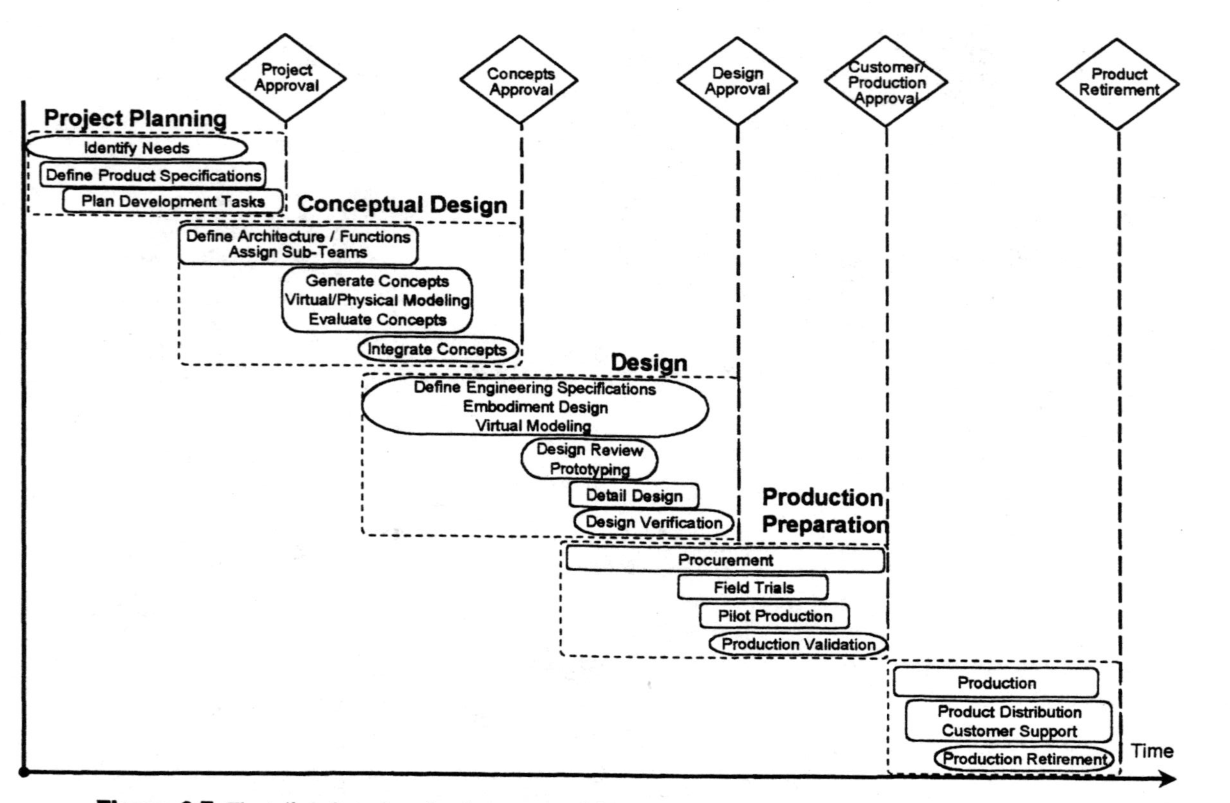

Figure 6.7 The tailored product development model for the example [Adapted with permission from H. Kemser].

Table 6.3 Assessment table for the process development model [Adapted with permission from J. Phair].

Section #	New Project Assessment Criteria		Level* A	B	C	D		Comments:
Planning								
1	Manufacturing Model Type	New					Minor	
Section #	New Project Assessment Criteria		Level* A	B	C			Comments:
Process Planning								
2	Manufacturing Process Complexity	High				Low		
3	Standards and Regulations	High				Low		
Product								
4	Product Design Paradigm	Original				Incremental		
5	Product Design Complexity	High				Low		
Sales and Marketing								
6	Time to Market	Short				Long		
7	Sales Volume	High				Low		
Resource Management								
8	Supplier Involvement	High				Low		
9	Partner Involvement	High				Low		
10	Procurement Needs	High				Low		

* *Evaluation of the Level depends on other products that are produced within the company.*

Table 6.4 Assessment table for the testing development model [Adapted with permission from J. Phair].

Section #	New Project Assessment Criteria		Level* A	B	C		Comments:
Test Planning							
1	Testing Design Type	New				Existing	
2	Testing Method Complexity	High				Low	
3	Standards and Regulations	High				Low	
Product							
4	Product Design Paradigm	Short				Long	
5	Product Design Complexity	Original				Incremental	
Sales and Marketing							
6	Time to Market	Short				Long	
7	Sales Volume	High				Low	
Resource Management							
8	Supplier Involvement	High				Low	
9	Partner Involvement	High				Low	
10	Procurement Needs	High				Low	

* *Evaluation of the Level depends on other products that are produced within the company.*

Table 6.5 Assessment table for the packaging development model [Adapted with permission from J. Phair]..

Section #	New Project Assessment Criteria		Level* A	B	C		Comments:
Packaging Planning							
1	Packaging Design Type	New				Existing	
2	Packaging Method Complexity	High				Low	
3	Standards and Regulations	High				Low	
Product							
4	Product Design Paradigm	Originial				Incremental	
5	Product Design Complexity	High				Low	
Sales and Marketing							
6	Time to Market	Short				Long	
7	Sales Volume	High				Low	
Resource Management							
8	Supplier Involvement	High				Low	
9	Partner Involvement	High				Low	
10	Procurement Needs	High				Low	

* *Evaluation of the Level depends on other products that are produced within the company.*

Planning

There are three criteria under the planning category. The first is the selection of the base manufacturing process model that will be tailored to fit project needs. The base models for process, testing, and packaging will be discussed below. The second criterion assesses the complexity of the project, and the third criterion assesses standards and regulation that govern the project.

Manufacturing Process Design Models As discussed in Chapter 5, there are three types of design models we are using in the manufacturing process design methodology: process design, testing design, and packaging design. We will begin by discussing the process design models.

The Process Development Models Four types of process design have been defined to reflect the types of situations encountered in manufacturing firms. Each of the four types begins with the project planning phase, which remains unchanged: identify needs, define product specifications, and plan development tasks. The next three phases differ between the models. The four models are shown in Figures 6.8-6.11. Figure 6.8 shows the most demanding of the four types of design situations, which is a new line using new technology. This type of design is usually accomplished in two parts. First, the new technology is brought in-house to be tested and to allow the engineers and designers to gain experience with it. This step occurs before the product design process begins. As stated several times in this book, a company should never develop a new product and a new manufacturing technology together. Too many unforeseen problems arise, delaying the product's delivery to market. Before applying a new manufacturing technology to a new product, preferably it should be applied to an existing product for testing. In general the first model shows the application of a new technology to an existing product, which will require a new line. This model includes all the design steps that were discussed in the last chapter. Figure 6.8 is the same as Figure 5.1 in the last chapter, but does not show the planning phase. The planning phase is not shown in any of the process models because it does not change from model to model.

The other three models have fewer design steps than those shown in Figure 6.8. As is the case in the product development models, not every design project requires the designers to complete every single design step, because some designs will require less effort, especially if using existing equipment. Therefore, we developed the other three models, which are adjusted to fit the particular situation. The model shown in Figure 6.9 is one for a new line using existing technology. This type of design project still requires a significant amount of effort because even though the technology is familiar, different machines may be used, a layout plan is still required, as is delivery, set-up, and debug. In the conceptual design phase, define engineering specifications and embodiment design have been combined into one step, as have generate concepts and design line layout. Finally, evaluate and integrate concepts have also been combined into a single step. As stated in the section on tailoring the product development model, when steps are combined, the outputs may be combined, but the team needs to ensure that the critical design information is still captured for future use and management team approval. The design phase and the production preparation phase remain unchanged from the previous model. Since this type of line design requires that a new line be installed, all the steps must be followed to ensure that the installation and validation go smoothly.

The model shown in Figure 6.10 is a design situation in which an existing manufacturing line is used for the new product, but the line will require major changes to accommodate the new product. The conceptual design phase is identical to that of the last model. Since the line will require major changes, it is anticipated that new stations will be designed and new machines will be purchased, thereby still requiring a significant design effort. In the design phase of this model, the steps define engineering specifications and embodiment design have been combined, as significant effort will be required, but a large portion of the line should still remain unchanged as far as layout is concerned. The rest of the steps remain separate in this model, as it is anticipated that new vendors may be chosen, facility requirements may change, the changes to the stations and line may require modeling and/or prototyping, and a line simulation should be per formed because of the changes. In the production preparation phase, procurement remains a separate step since new equipment is anticipated for this line. However, since there may not be a large number of new machines, installation, equipment integration, and equipment acceptance are combined into a single step. Debug,

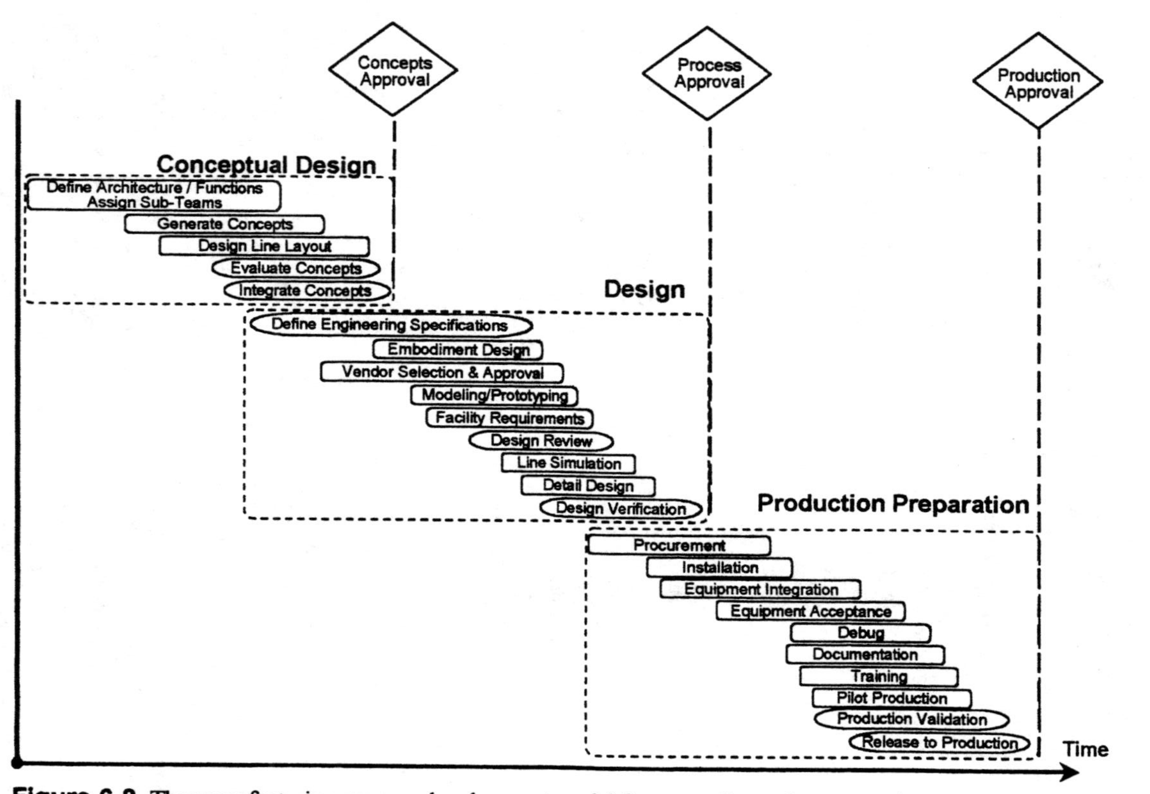

Figure 6.8 The manufacturing process development model for a new line using new technology [Adapted with permission from J. Phair].

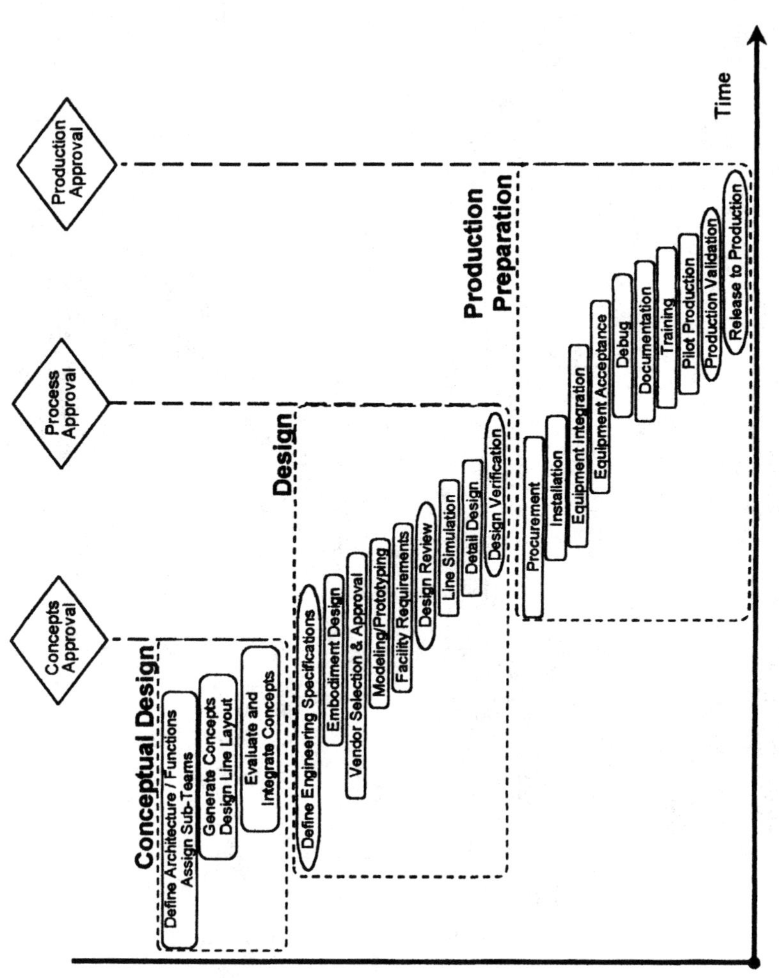

Figure 6.9 The manufacturing process development model for a new line using existing technology [Adapted with permission from J. Phiar].

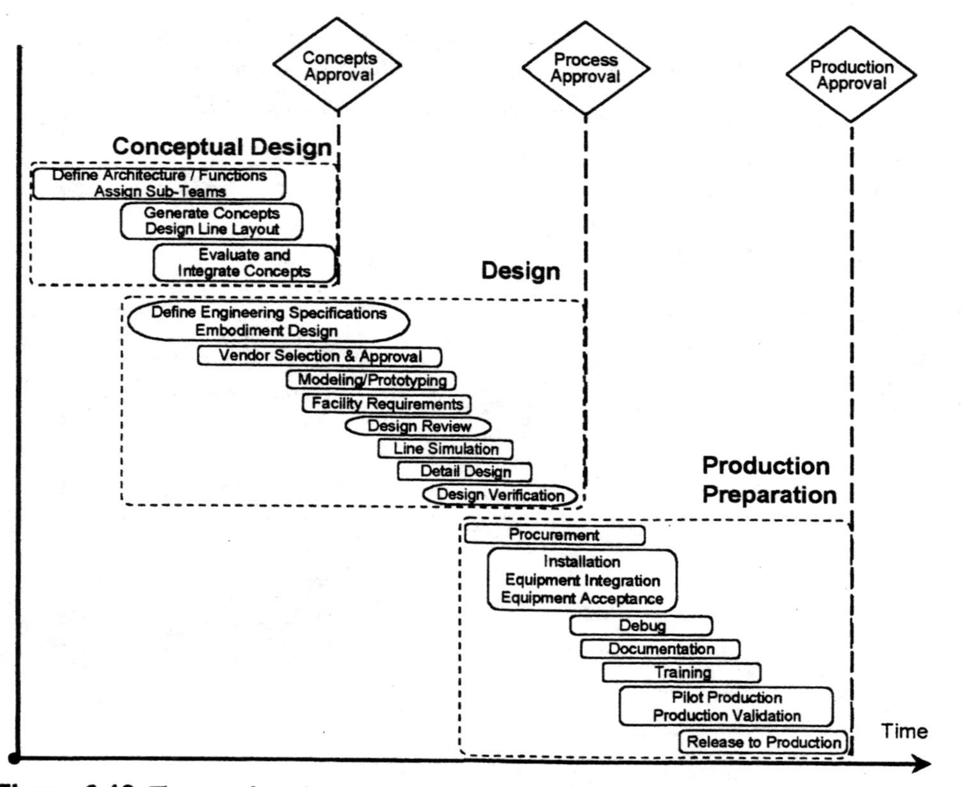

Figure 6.10 The manufacturing process development model for an existing line needing major modifications [Adapted with permission from J. Phiar].

documentation, and training remain separate steps because of their importance to the success of the manufacturing line. Pilot production and production validation have been combined, as for the most part the line remains intact from previous products. Finally, the step, release to production, remains separate, as it is a formal hand-off of responsibility to the product team.

The last model type for the manufacturing process development is the case in which there are minor modifications to an existing line. This model type is shown in Figure 6.11. In the conceptual design phase of this model, there is only one step, generate and evaluate concepts, because few changes to the line are expected. In the design phase, there are three steps: 1) define engineering specifications and embodiment design, 2) design review, line simulation, and detail design, and 3) design verification. The design steps, vendor selection, facility requirements, and modeling/prototyping have been eliminated because all changes to the line are expected to be handled in-house. In the production preparation phase, the steps procurement and equipment acceptance have been eliminated as no equipment is expected to be purchased for this line. Installation and integration have been combined to reflect the amount of modification anticipated. However, debug, documentation, and training remain separate steps because of their importance to the success of the manufacturing line. pilot production and production validation have been combined, as for the most part the line remains intact from previous products. Finally, as in the model above, the step, release to production, remains separate, as it is a formal hand-off of responsibility to the product team.

The Test Development Models There are three model types for test development: 1) the test methods in which new technologies are used, 2) the test methods in which existing technologies are used, and 3) the test methods in which existing equipment is used. Figure 6.12 shows the first model in which new technologies are chosen for testing the new product. This is the same model as that shown in Figure 5.2, therefore we will not discuss this model here.

Figure 6.13 shows the model for using existing technology, which is technology that has been previously used within the company for testing the new product. In the conceptual design phase, all the steps remain separate except evaluate and integrate concepts. This step has been combined as these steps should be fairly straightforward because of the familiarity of the technology. The other steps have remained separate, because new machines may be considered with the new product demands. The design phase remains unchanged from the last model, as does the production preparation phase, because new machines will be brought in for testing the new product. Therefore, all the steps are important for a well functioning test system(s).

Figure 6.14 shows the model in which existing equipment is modified to test the new product. In the conceptual design phase of this model, there are two steps, testing requirements and generate and evaluate concepts. Because product testing is such an important part of the manufacturing process, testing requirements need

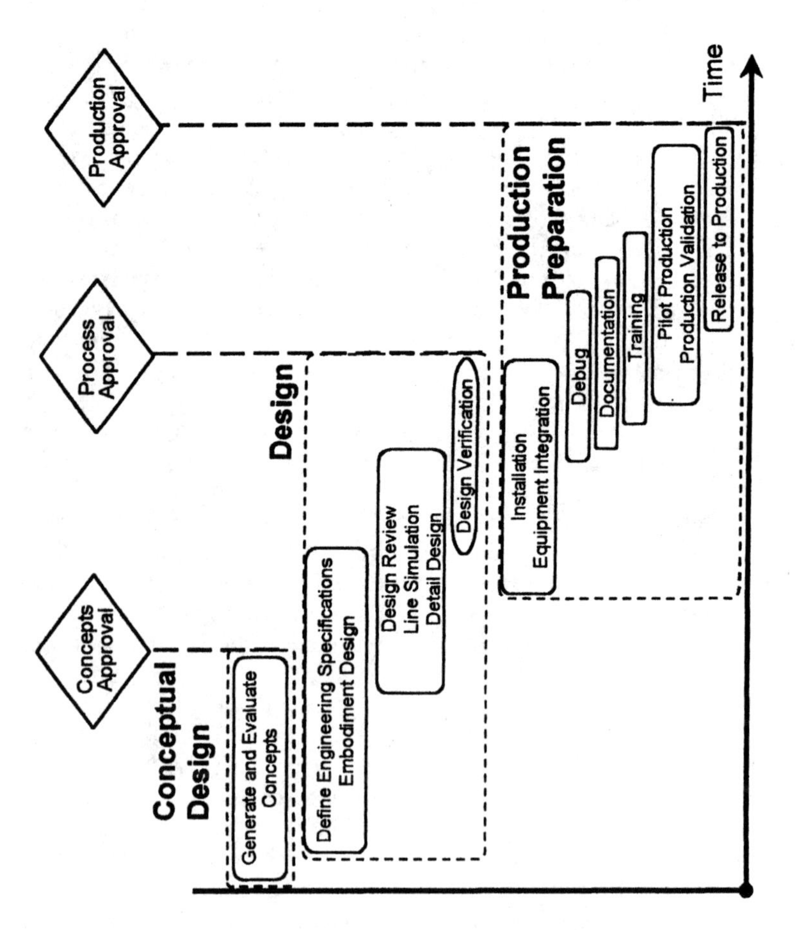

Figure 6.11 The manufacturing process development model for an existing line needing minor modifications [Adapted with permission of J. Phair].

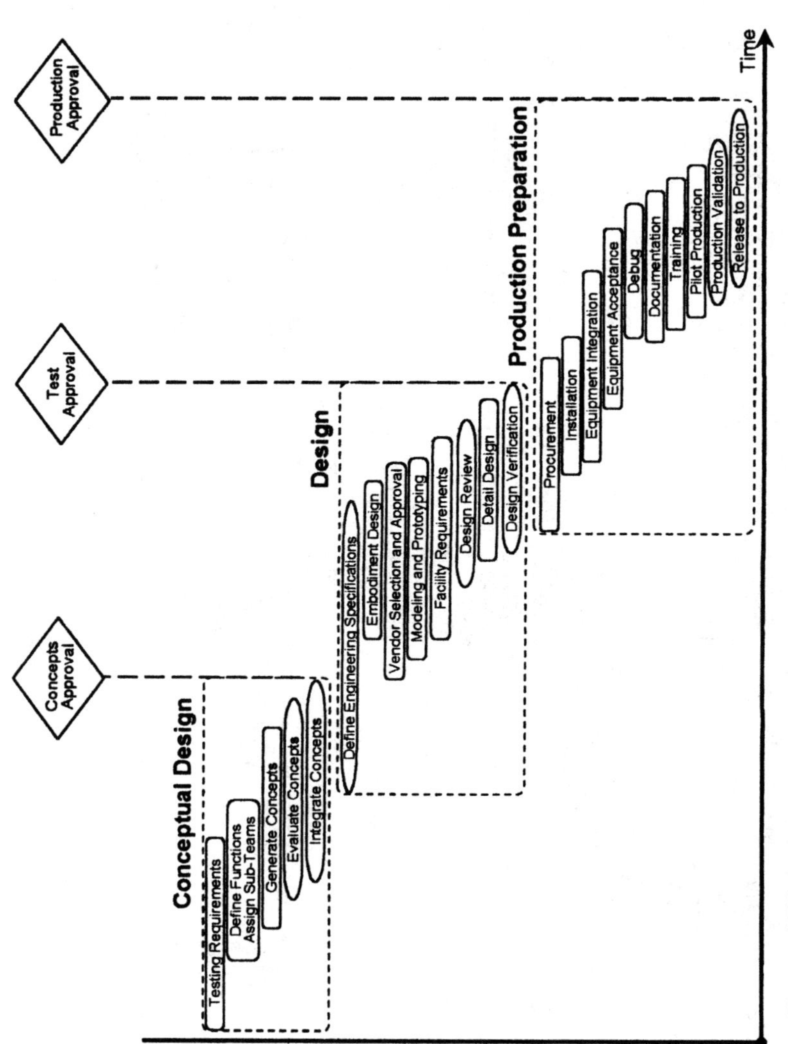

Figure 6.12 The test development model in which new technology is used for testing the new product [Adapted with permission from J. Phair].

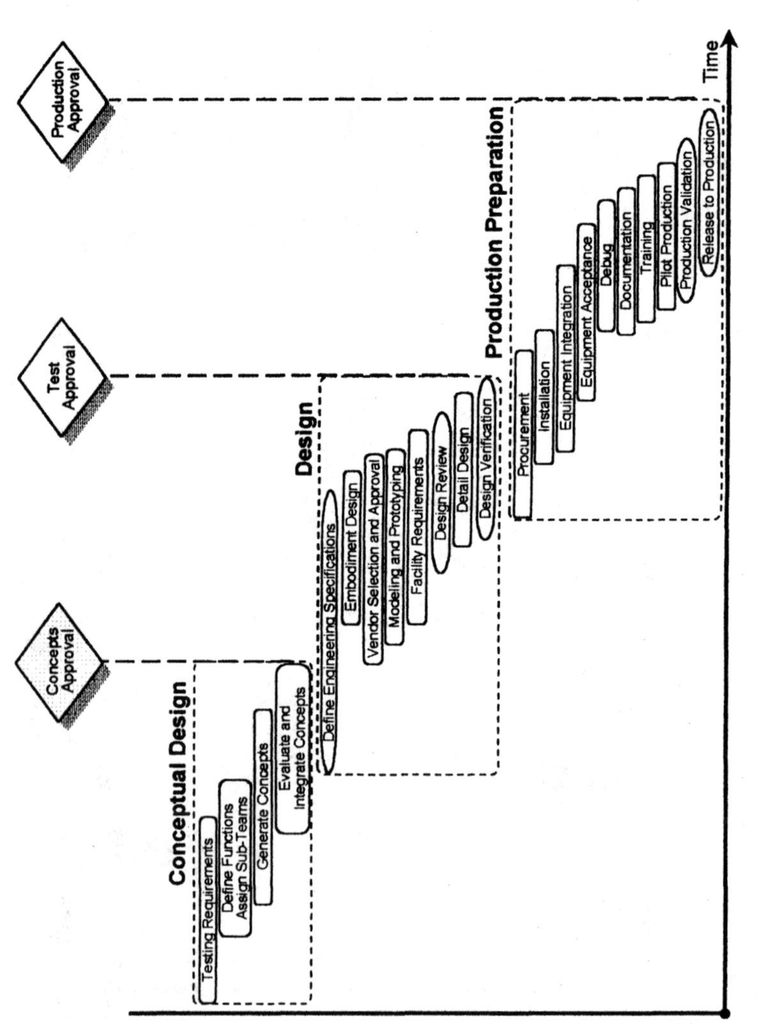

Figure 6.13 The test development model in which an existing technology is used for product testing [Adapted with permission from J. Phair].

to be considered carefully. The next step is combined, as only minor changes are expected in the existing equipment. In the design phase, there are three steps: 1) define engineering specifications and embodiment design, and facility requirements 2) design review and detail design, and 3) design verification. The vendor selection and modeling/prototyping design steps were eliminated because all changes to the line are expected to be handled in-house. Facility requirements remain a part of the phase as new testing requirements for this product may require different electrical needs. In the production preparation phase, the procurement and equipment acceptance steps were eliminated because no equipment is expected to be purchased for this line. Installation and integration were combined to reflect the amount of modification anticipated. However, debug, documentation, and training remain separate steps because of their importance to the success of the manufacturing line and are needed for checking the test process and ensuring personnel can run the equipment. Pilot production and production validation have been combined. Finally, as in the model above, the release to production step remains separate, since it is a formal hand-off of responsibility to the product team.

The Packaging Development Models There are three model types for packaging development: 1) the packaging methods in which new technologies are used, 2) the packaging methods in which existing technologies are used, and 3) the packaging methods in which existing equipment is used. Figure 6.15 shows the first model in which new technologies are chosen for packaging the new product. This is the same model as shown in Figure 5.3; therefore, the discussion of this model will be skipped.

Figure 6.16 shows the model for using existing technology, which is technology that has been previously used within the company, for packaging the new product. In the conceptual design phase of packaging requirements, all of the steps remain separate except evaluate and integrate concepts. This step has been combined as these steps should be fairly straightforward because of the familiarity of the technology. The other steps have remained separate, because new machines may be considered with the new product demands. The design phase remains unchanged from the last model, as does the production preparation phase, because new machines will be brought in for packaging the new product. Therefore, all the steps are important for a packaging line that can package the product for safe shipment.

Figure 6.17 shows the model in which existing equipment is modified for packaging the new product. In the conceptual design phase of this model, there are two steps: packaging requirements, and assign sub-teams, generate and evaluate concepts. Because packaging is important to ensure that the product arrives in good working condition at the customer site, packaging requirements need to be carefully considered. The next step is combined, as only minor changes are expected in existing equipment, however, sub-teams may be needed depending on

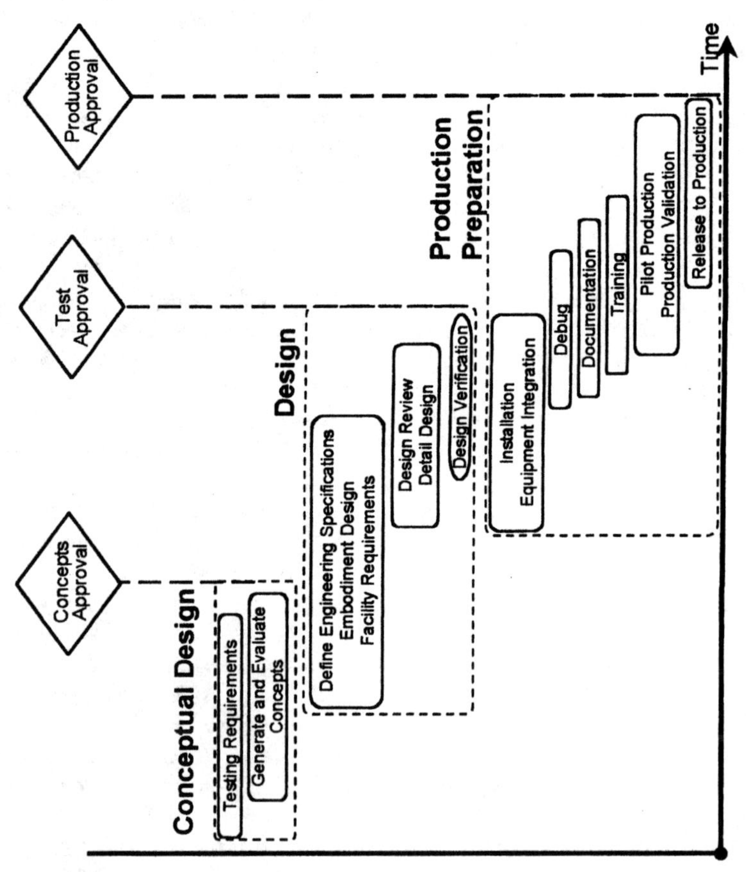

Figure 6.14 The test development model that uses existing equipment [Adapted with permission from J. Phair].

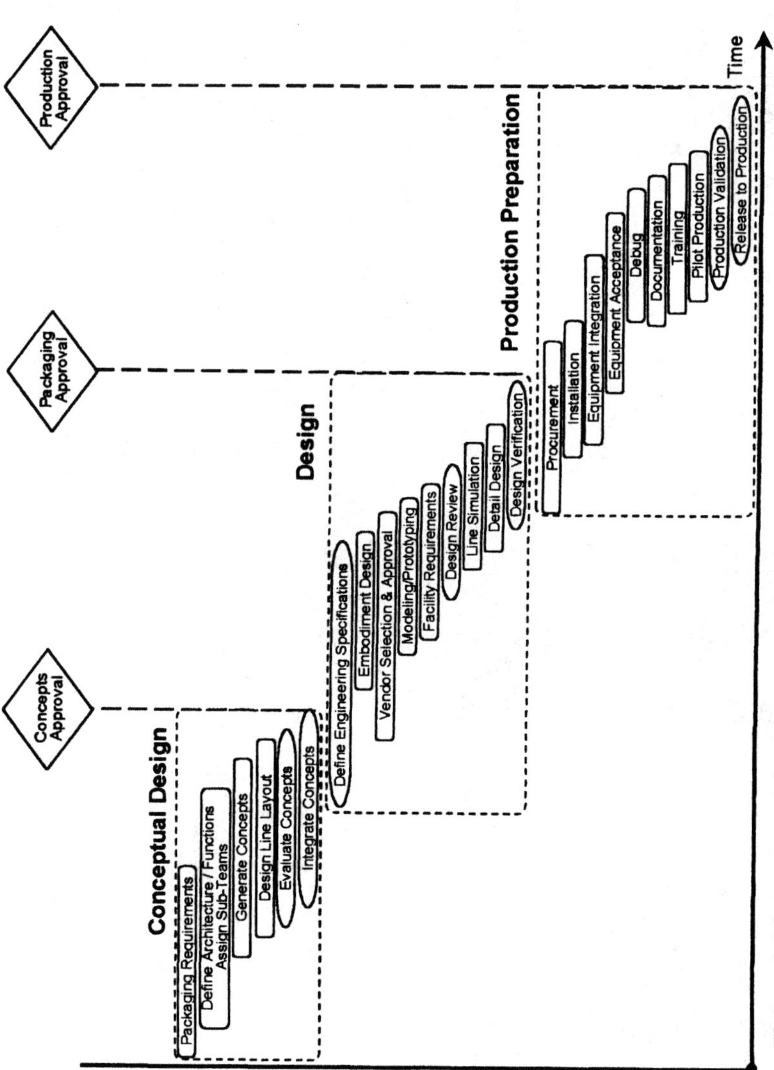

Figure 6.15 The packaging development model in which new technology is used for packaging the new product [Adapted with permission from J. Phair].

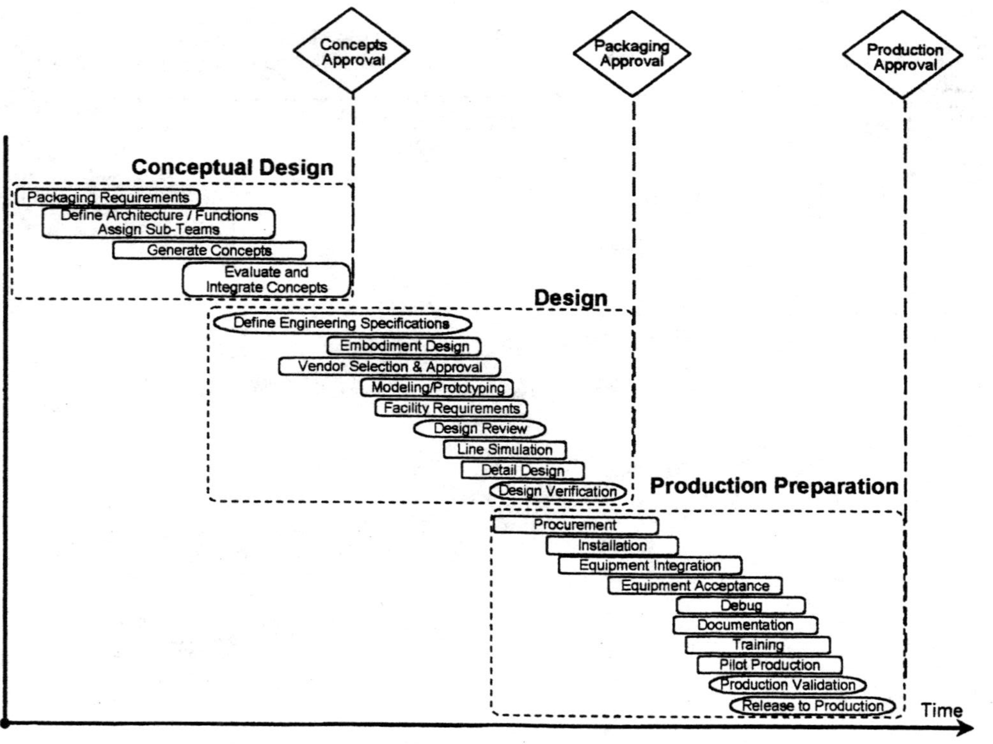

Figure 6.16 The packaging development model in which existing technology is used for packaging the new product [Adapted with permission from j. Phair].

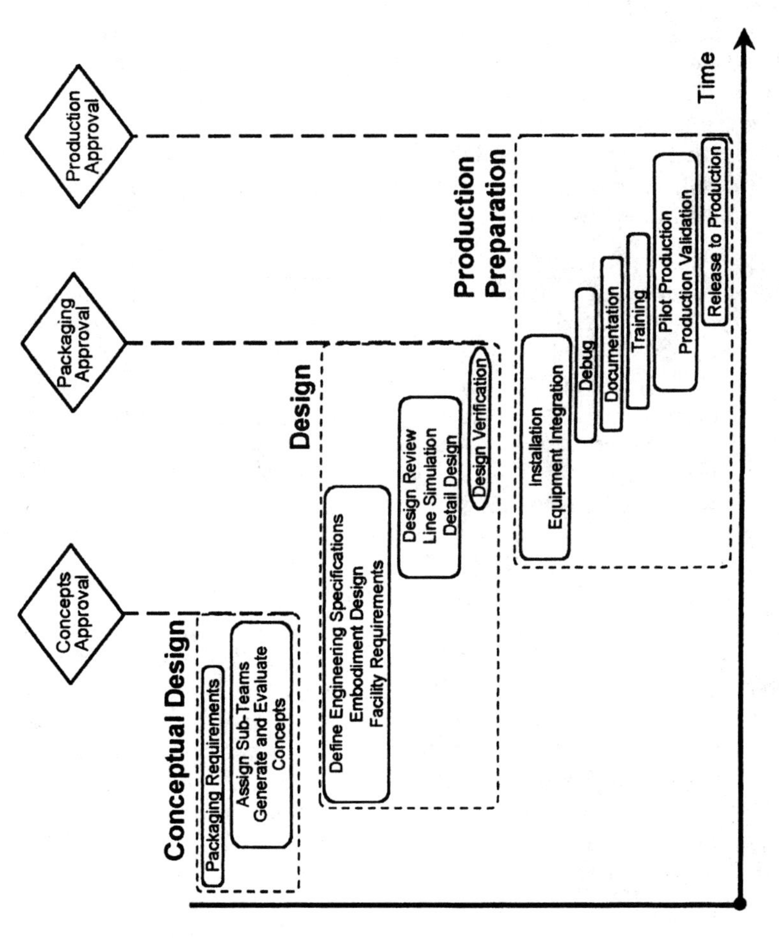

Figure 6.17 The packaging development model in which existing equipment is used for packaging the new product [Adapted with permission from J. Phair].

the complexity of the packaging line. In the design phase, there are three steps: 1) define engineering specifications and embodiment design, and facility requirements 2) design review and detail design, and 3) design verification. The vendor selection and modeling/prototyping design steps were eliminated because all changes to the line are expected to be handled in-house. Facility requirements remains a part of the phase as the equipment may have different needs from previous products. In the production preparation phase, the steps procurement and equipment acceptance were eliminated as no equipment is expected to be purchased for this line. Installation and integration were combined to reflect the amount of modification anticipated. However, debug, documentation, and training remain separate steps because of their importance to the success of the manufacturing line and the need for safely shipping the product which relies on the packaging. Pilot production and production validation have been combined. Finally, as in the model above, the release to production step remains separate, as it is a formal hand-off of responsibility to the product team.

Complexity The second criterion to be evaluated is complexity, which is complexity of the manufacturing process, of the testing method, or of the packaging method, depending on which model we are addressing. In any case, we define complexity as a measure of the skill level of the personnel needed for the project, which is decided by the types of technology used and in some measure the number of stations required by the line to accomplish the task, the number of team members needed to design the process, the development hours estimated, and the level of integration needed between teams. All these issues help the team decide whether the project is of high, medium, or low complexity. As stated in the product section, complexity must be evaluated within the context of other products and projects within the company itself. What is high complexity for one company, may be low complexity for another. If a company is undertaking a new process such as injection molding, which requires new skills and coordination with the product team and suppliers, then that process would be rated as having high complexity. If on the other hand, a company currently uses injection molding, but a new complex mold is required in the manufacturing process, then the project would be rated as having average complexity. Finally, a low complexity process would be one in which the same mold from another project or one requiring only minor changes would be used. The same examples would hold true for the testing and packaging development models.

Standards and Regulations The last criterion to be evaluated in this category is standards and regulations. There are several types of standards that might need to be addressed including government regulations such as OSHA rules, international and national standards requirements such as ISO quality standards, and internal company standards. In addition, the customer may also impose rigid standards that must be met and verified through testing. As discussed before, evaluating this category is very company-dependent. What one company may consider to be rigid requirements may be typical requirements to another. If the standards and regulations imposed on the new project are significantly higher or lower than other projects within the company, then the rating would be high (Level A) or low (Level C), respectively. Again, this is a judgment call made by the team.

Product

There are two criteria under the product category. The first is the product design paradigm that was determined by the CPDT. The second is the complexity of that design as determined by the CDT.

Product Design Paradigm The product design paradigm gives the CPDT an indication of how much interaction there needs to be with the CDT. If the design is defined as original, then it is new to the company and will require more interaction with the product team, and will require significant testing and validation to ensure that the product made on the manufacturing line meets all the product specifications and requirements. If the design is evolutionary, then the company personnel have more familiarity with the design, and less interaction is needed between the teams. Finally, if the design is incremental, then the manufacture of the product should be a straightforward task, as all teams have a high degree of familiarity with the design.

Product Design Complexity This criterion was determined by the CDT in their evaluation of the product. It indicates to the CPDT the level of testing and validation that will be needed on the manufacturing line to ensure that the complex product that is produced on the line meets all product specifications. A complex product does not necessarily indicate that the process, testing or packaging methods are rated as complex. However, it does indicate the need for validation of the product on these lines.

Sales and Marketing

There are two criteria considered in this category: time-to-market and sales volume. Again, both of these criteria were addressed by the CDT in the product assessment. The same answers should be applied in these assessments. The time-to-market affects the scheduling of the tasks and the number of personnel needed on the design project. A short time will require more overlap of design steps and more personnel, and a long time-to-market will require less overlap and fewer personnel. The second criterion, sales volume, affects the amount of testing and validation needed on the line. A line that is expected to produce high volumes will need more testing and validation since it will be expected to perform at a high rate of production for a significant portion of the week. On the other hand, a line that produces a low number of products may not be tested thoroughly, since few products will be run down the line, and these few products may be too expensive to produce in a test situation.

Resource Management

There are three criteria considered in this category: supplier involvement, partner involvement, and procurement needs. How much equipment and parts the suppliers provide to the manufacturing line can have an impact on team interactions. Suppliers supply the company with parts or equipment, but the role of a partner is much greater. A partner can either manufacture part of the new product either at the company or at the partner site, or a partner can design equipment for the manufacturing, testing, or packaging lines. A high level of involvement (Level A) would indicate that the third party is part of all major decisions throughout the development of the equipment for the line, and therefore is a part of the team as indicated by the support personnel, or in the case of the partner, may actually be a team member of the CPDT. A low level of involvement (Level C) indicates that there is no involvement by the third party or it comes very late in the process, such as in pilot production.

Procurement needs gives an indication as to how much equipment and/or parts are purchased for use on the manufacturing and packaging lines. This criterion affects whether a design step is needed for vendor selection, procurement, and equipment acceptance.

The project assessment tool for the manufacturing process, testing, and packaging models need to be undertaken with support from the CDT, as the answers that they supplied for the product design will carry over into these models, as indicated above. The first time the CPDT uses the project assessment tool will be time consuming, since they need to develop guidelines for its use. These guidelines will be a starting point for deciding what is considered average within the company, and may change as more products are developed using this methodology. However, iteration is to be expected in concurrent engineering and continuous process improvement is an important part of this methodology.

6.2.2 Manufacturing Process Development Targets

Once the project assessments have been completed for the three models, the results are plotted on a bull's-eye graph as shown in Figures 6.18-6.19. The graph is divided into the four categories discussed above: planning, product, sales and marketing, and resource management. These categories are further divided into sections that correspond to the evaluation criteria discussed above. There are three rings on the graph, A, B, and C, that correspond to the levels in the evaluation table, with A being the outermost ring, and C being the innermost. Figure 6.18, the process development target, contains a section for the criterion on the type of process design with four levels to this criterion corresponding to the four models. Figure 6.19 represents the target for the test and packaging methods.

The goal of the development team is to reach the center of the target—a new manufacturing process line, a new testing method, and a new packaging method or line. The plotted evaluation criteria give an indication of the effort and resources required to reach the goal of a new design by their distance from the center.

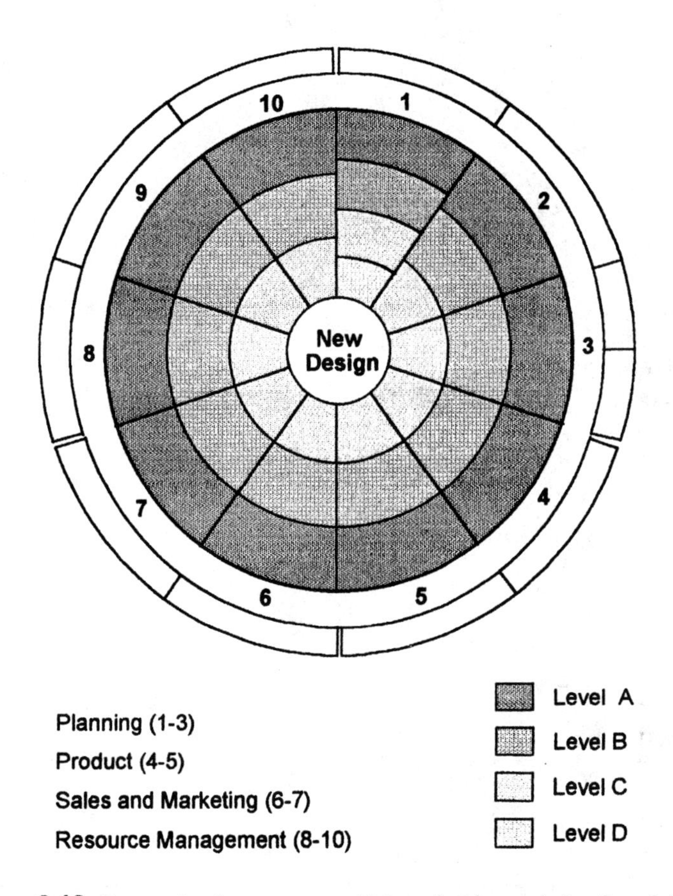

Figure 6.18 Process development target [Adapted with permission from J. Phair].

These plotted points will now be used to create a process, testing and/or packaging development model for this project. The following sections describe the steps to create these models.

6.2.3 Creating the Customized Manufacturing Process Development Models

Using the answers from Tables 6.3-6.5, the three manufacturing process models can be customized to fit the needs of the development project. Each criterion has an effect on the model, and will either streamline the steps or expand them. First, the major milestones of the project are defined. Each project begins with a kick-

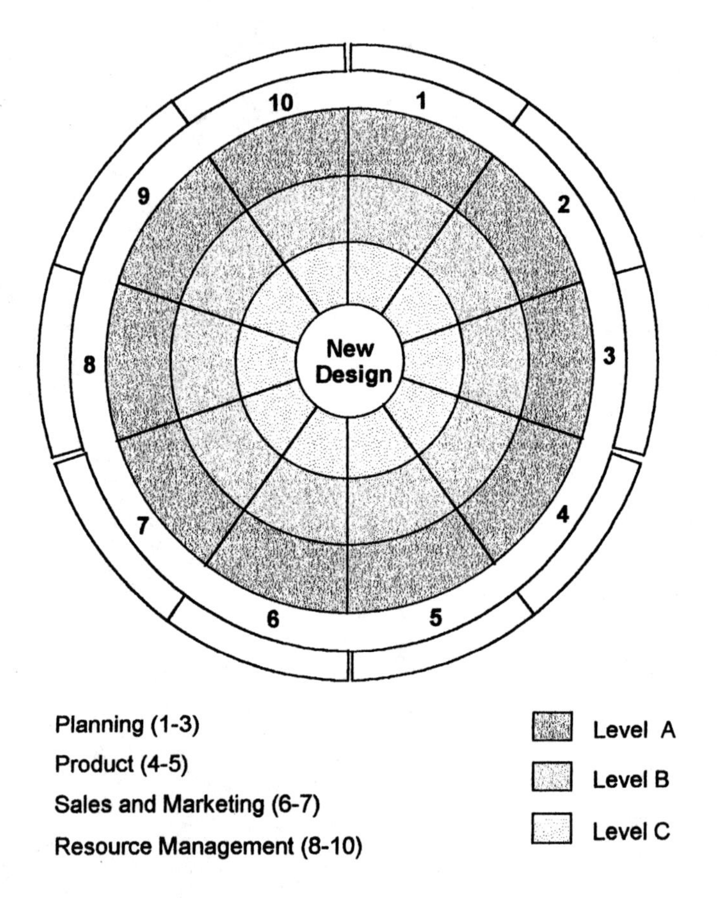

Figure 6.19 Development target for testing and packaging models [Adapted with permission from J. Phair].

off meeting, followed by a planning phase, which always consists of three design steps: identify needs, define product specifications, and plan development tasks. The design process ends with the production and service phase, which was discussed in the last chapter. We will now look at how each of the criteria affect the three models.

The Process Model

The process model has ten criteria which can be tailored to the project. Each will be discussed as in the section on the product model.

Section 1: Manufacturing Model Type
Level A (New Line and New Technology) The model shown in Figure 6.8 is used as the base model for this project.
Level B (New Line and Existing Technology) The model shown in Figure 6.9 is used as the base model for this project.
Level C (Existing Line, Major Changes) The model shown in Figure 6.10 is used as the base model for this project.
Level D (Existing Line, Minor Changes) The model shown in Figure 6.11 is used as the base model for this project.

Section 2: Manufacturing Process Complexity
This criterion affects conceptual design and the design phases. It also indicates a need for multiple development teams, and the level of expertise needed within the teams.
Level A (High) Separate development steps are created for the definition of the product architecture/functions and assign sub-teams, generate concepts, integrate concepts, modeling/prototyping, debug, design review, line simulation, and design verification. An additional multi-disciplinary product team should be created and additional personnel with critical skills are gathered to meet the needs of the project.
Level B (Average) The development steps concept integration and modeling/prototyping should be included in the model.
Level C (Low) No changes to the model are required.

Section 3: Standards and Regulations
This criterion affects the conceptual design and design phases, and may call for additional support team members.
Level A (High) Separate development steps are created for evaluate concepts, modeling/prototyping, and design review. The CDT should be a support member for all these steps.
Level B (Average) The integrate concepts development step should be included in the model. The CDT should be included as a support team member for this step.
Level C (Low) No effect on the model.

Section 4: Product Design Paradigm
This criterion affects the conceptual design and production preparation phases, and the level of integration between the product and process development teams.
Level A (Original) Separate development steps are created for generate concepts, line layout, evaluate concepts, integrate concepts, pilot production, and production validation. The CDT should be a support team member in all these steps.
Level B (Evolutionary) The production validation development step should be included in the model with the CDT as a support team member.
Level C (Incremental) No effect on the model.

Section 5: Product Design Complexity
This criterion affects the conceptual design and production preparation phases and team integration.
Level A (High) Separate development steps are created for generate concepts, line layout, modeling/prototyping, debug, and production validation. The CDT should be a support team member for these steps.
Level B (Average) The development step production validation should be included in the model with the CDT as a support team member.
Level C (Low) No changes to the model are required.

Section 6: Time-to-Market
This criterion affects the overlap and integration of the design activities. The shorter the time-to-market, then the more of the tasks need to be overlapped to meet the deadlines imposed. However, note that the more the tasks are overlapped the higher the risk is for cost overruns and problems.
Level A (Short) The delivery date is set by marketing in conjunction with the CDT and CPDT with approval by the management team. From this date, the critical tasks are scheduled backwards, and more personnel are added to ensure that this date can be met.
Level B (Average) No changes to the model are required.
Level C (Long) There is little overlap of the development steps, and there is little undertaking of risk on the project. There is little or no parallel development in this type of project.

Section 7: Sales Volume
This criterion affects the design and production preparation phases.
Level A (High) Separate development steps are created for line simulation, pilot production, and production validation.
Level B (Average) There is no effect on the model.
Level C (Low) Pilot production is eliminated from the model.

Section 8: Supplier Involvement
Level A (High) The supplier should be included as a support team member in the following steps: generate concepts, evaluate concepts, integrate concepts, embodiment design, design review, installation, equipment integration, equipment acceptance, debug, pilot production, and production validation.
Level B (Average) The supplier should be included as a support team member during generate concepts, evaluate concepts, embodiment design, modeling/prototyping, equipment acceptance, and pilot production.
Level C (Low) No effect on the model.

Section 9: Partner Involvement
A partner may be involved in creating critical stations or components, creating some stations or components, or have no involvement at all.

Level A (High) The partner should have personnel as members of the CPDT.
Level B (Average) The partner should have personnel as support team members of the CPDT.
Level C (Low) No effect on the model.

Section 10: Procurement Needs
Level A (High) Separate steps should be created for the following: vendor selection and approval, procurement, equipment acceptance, installation, equipment integration, and debug.
Level B (Average) The following steps are needed in the model, although they do not need to be separate if they are already in the model: vendor selection and approval, procurement, equipment acceptance, installation, equipment integration, and debug.
Level C (Low) No effect on the model or if the steps vendor selection and procurement are in the model, they may be removed.

The Test Model

The test process has ten criteria used for tailoring the selected model. Each one will be discussed individually.

Section 1: Test Model Type
Level A (New Line and New Technology) The model shown in Figure 6.11 is used as the base model for this project.
Level B (New Line and Existing Technology) The model shown in Figure 6.12 is used as the base model for this project.
Level C (Existing Equipment) The model shown in Figure 6.13 is used as the base model for this project.

Section 2: Test Method Complexity
This criterion affects the conceptual design, design, and production preparation phases. It also indicates the level of expertise needed within the teams.
Level A (High) Separate development steps are created for define functions and assign sub-teams, generate concepts, integrate concepts, modeling/prototyping, design review, design verification, and debug. Additional personnel with critical skills are gathered to meet the needs of the project.
Level B (Average) The integrate concepts and modeling/prototyping development steps should be included in the model.
Level C (Low) No changes to the model are required.

Section 3: Standards and Regulations
This criterion affects the conceptual design and design phases, and may call for additional support team members.

Level A (High) Separate development steps are created for evaluate concepts, modeling/prototyping, and design review. The CDT should be a support member for all of these steps.
Level B (Average) The integrate concepts development step should be included in the model. The CDT should be included as a support team member for this step.
Level C (Low) No effect on the model.

Section 4: Product Design Paradigm
This criterion affects the conceptual design and production preparation phases, and the level of integration between the product and process development teams.
Level A (Original) Separate development steps are created for define functions and assign sub-teams, generate concepts, evaluate concepts, integrate concepts, pilot production, and production validation. The CDT should be a support team member in all these steps.
Level B (Evolutionary) The production validation development step should be included in the model with the CDT as a support team member.
Level C (Incremental) No effect on the model.

Section 5: Product Design Complexity
This criterion affects the conceptual design and production preparation phases and team integration.
Level A (High) Separate development steps are created for generate concepts, modeling/prototyping, debug, and production validation. The CDT should be a support team member for these steps.
Level B (Average) The production validation development step should be included in the model with the CDT as a support team member.
Level C (Low) No changes to the model are required.

Section 6: Time-to-Market
This criterion affects the overlap and integration of the design activities. The shorter the time-to-market, then the more tasks need to be overlapped to meet the deadlines imposed. However, note that the more the tasks are overlapped the higher the risk is for cost overruns and problems.
Level A (Short) The delivery date is set by marketing in conjunction with the CDT and CPDT with approval by the management team. From this date, the critical tasks are scheduled backwards, and more personnel are added to ensure that this date can be met.
Level B (Average) No changes to the model are required.
Level C (Long) There is little overlap of the development steps, and there is little undertaking of risk on the project. There is little or no parallel development in this type of project.

Section 7: Sales Volume
This criterion affects the design and production preparation phase.
Level A (High) Separate development steps are created for pilot production and production validation.
Level B (Average) There is no effect on the model.
Level C (Low) Pilot production is eliminated from the model.

Section 8: Supplier Involvement
Level A (High) The supplier should be included as a support team member in the following steps: generate concepts, evaluate concepts, integrate concepts, embodiment design, design review, installation, equipment integration, equipment acceptance, debug, pilot production, and production validation.
Level B (Average) The supplier should be included as a support team member during generate concepts, evaluate concepts, embodiment design, modeling/prototyping, equipment acceptance, and pilot production.
Level C (Low) No effect on the model.

Section 9: Partner Involvement
A partner may be involved in creating critical stations or components, creating some stations or components, or have no involvement at all.
Level A (High) The partner should have personnel as members of the CPDT.
Level B (Average) The partner should have personnel as support team members of the CPDT.
Level C (Low) No effect on the model.

Section 10: Procurement Needs
Level A (High) Separate steps should be created for the following: vendor selection and approval, procurement, installation, equipment acceptance, equipment integration, and debug.
Level B (Average) The following steps are needed in the model (although they do not need to be separate if they are already in the model): vendor selection and approval, procurement, installation, equipment acceptance, equipment integration, and debug.
Level C (Low) No effect on the model or if the vendor selection and procurement steps are in the model, they may be removed.

The Packaging Model

The packaging process has ten criteria used for tailoring the selected model. Each one will be discussed individually.
Section 1: Packaging Model Type
Level A (New Technology) The model shown in Figure 6.15 is used as the base model for this project.
Level B (Existing Technology) The model shown in Figure 6.16 is used as the base model for this project.

Level C (Existing Equipment) The model shown in Figure 6.17 is used as the base model for this project.

Section 2: Packaging Method Complexity

This criterion affects the conceptual design, design, and production preparation phases. It also indicates a need for additional personnel, and the level of expertise needed within the teams.

Level A (High) Separate development steps are created for the define architecture/functions and assign sub-teams, generate concepts, integrate concepts, modeling/prototyping, design review, line simulation, design verification, and debug. Additional personnel with critical skills are added to meet the needs of the project.

Level B (Average) The integrate concepts and modeling/prototyping development steps should be included in the model.

Level C (Low) No changes to the model are required.

Section 3: Standards and Regulations

This criterion affects the conceptual design and design phases, and may call for additional support team members.

Level A (High) Separate development steps are created for evaluate concepts, modeling/prototyping, and design review. The CDT should be a support member for all these steps.

Level B (Average) The integrate concepts development step should be included in the model. The CDT should be included as a support team member for this step.

Level C (Low) No effect on the model.

Section 4: Product Design Paradigm

This criterion affects the conceptual design and production preparation phases, and the level of integration between the product and process development teams.

Level A (Original) Separate development steps are created for generate concepts, design line layout, evaluate concepts, integrate concepts, pilot production, and production validation. The CDT should be a support team member in all these steps.

Level B (Evolutionary) The production validation development step should be included in the model with the CDT as a support team member.

Level C (Incremental) No effect on the model.

Section 5: Product Design Complexity

This criterion affects the conceptual design and production preparation phases and team integration.

Level A (High) Separate development steps are created for generate concepts, design line layout, modeling/prototyping, debug, and production validation. The CDT should be a support team member for these steps.

Level B (Average) The development step production validation should be included in the model with the CDT as a support team member.
Level C (Low) No changes to the model are required.

Section 6: Time-to-Market
This criterion affects the overlap and integration of the design activities. The shorter the time-to-market, then the more the tasks need to be overlapped to meet the deadlines imposed. However, note that the more the tasks are overlapped the higher the risk is for cost overruns and problems.
Level A (Short) The delivery date is set by marketing in conjunction with the CDT and CPDT with approval by the management team. From this date, the critical tasks are scheduled backwards, and more personnel are added to ensure that this date can be met.
Level B (Average) No changes are required of the model.
Level C (Long) There is little overlap of the development steps, and there is little undertaking of risk on the project. There is little or no parallel development in this type of project.

Section 7: Sales Volume
This criterion affects the design and production preparation phases.
Level A (High) Separate development steps are created for line simulation, pilot production, and production validation.
Level B (Average) There is no effect on the model.
Level C (Low) Pilot production is eliminated from the model.

Section 8: Supplier Involvement
Level A (High) The supplier should be included as a support team member in the following steps: generate concepts, evaluate concepts, integrate concepts, embodiment design, design review, installation, equipment integration, equipment acceptance, debug, pilot production, and production validation.
Level B (Average) The supplier should be included as a support team member during generate concepts, evaluate concepts, embodiment design, modeling/prototyping, equipment acceptance, and pilot production.
Level C (Low) No effect on the model.

Section 9: Partner Involvement
A partner may be involved in creating critical stations or components, creating some stations or components, or have no involvement at all.
Level A (High) The partner should have personnel as members of the CPDT.
Level B (Average) The partner should have personnel as support team members of the CPDT.
Level C (Low) No effect on the model.

Section 10: Procurement Needs
Level A (High) Separate steps should be created for the following: vendor selection and approval, procurement, equipment acceptance, installation, equipment integration, and debug.
Level B (Average) The following steps are needed in the model (although they do not need to be separate if they are already in the model): vendor selection and approval, procurement, equipment acceptance, installation, equipment integration, and debug.
Level C (Low) No effect on the model or if the vendor selection and procurement steps are in the model, they may be removed.

6.2.4 Customizing the Manufacturing Process Development Models

This is the final step of the tailoring process. The purpose of this step is to streamline or expand the base models to ensure that all the needed development steps are included. Once the model is complete, no steps should be removed, because their removal could lead to a loss of information within the design process, or may require unnecessary iteration. The next section will show how this tailoring process is accomplished using a design example from a small company.

6.2.5 Tailoring Process Example

A safe company wants to add a new feature to their safes that will improve the product's versatility, namely adjustable interior shelving. The new shelving will require more and different work to be done in some of the manufacturing stations. Most of the labor in manufacturing the safes is manual, with a few tasks aided by machines. To change from a fixed-shelf configuration to an adjustable shelf design, a series of rails need to be installed on the sides and back of the safes. The shelves are then hung on the rails using special hangers.

First, the company must clearly identify the needs of the project. The design should allow for shelving that will hold heavy objects safely and securely, the shelves should be easy to use and adjust, and the new design should not compromise the quality of the safe. The company wants to run the new design down the existing manufacturing line, making only those adjustments in the line that need to be made to accommodate the new shelving units.

Next, the management team and the CPDT need to assess the project using the assessment tables shown in Tables 6.3-6.5. Since no testing or packaging changes will be needed to accommodate the new design, only Table 6.3, the process assessment table is needed for this project. Table 6.6 shows the assessment for this new project.

Looking at the assessment table, we see that the manufacturing model chosen was "minor changes to an existing line", discussed above. For the process planning category, the manufacturing process complexity was rated as low,

Table 6.6 Assessment table for the safe example [Adapted with permission from J. Phair].

Section #	New Project Assessment Criteria		Level* A	B	C	D		Comments:
Planning								
1	Manufacturing Model Type	New				X	Minor	
Section #	New Project Assessment Criteria		Level* A	B	C			Comments:
Process Planning								
2	Manufacturing Process Complexity	High			X	Low		
3	Standards and Regulations	High			X	Low		
Product								
4	Product Design Paradigm	Original			X	Incremental		
5	Product Design Complexity	High			X	Low		
Sales and Marketing								
6	Time to Market	Short	X			Long		Introduction driven by competition
7	Sales Volume	High		X		Low		Expect average sales
Resource Management								
8	Supplier Involvement	High			X	Low		Supplier for hangers
9	Partner Involvement	High			X	Low		
10	Procurement Needs	High			X	Low		

** Evaluation of the Level depends on other products that are produced within the company.*

as was standards and regulations. Only minor changes are made to the design, thereby having little impact on complexity and little on standards, except to ensure that the product meets company standards. In the product category, both criteria were rated as low. The product design change is incremental, and the design complexity is low, as it is a matter of adding rails and hangers to allow for shelf adjustment. In the sales and marketing category, the time-to-market is short, as competitors are already offering this option on their safes, and the company is losing sales without this new feature. The sales volume for this product is expected to be about the same as their other safe lines. In the resource management category, there is no partner involvement. However, a supplier is needed for the hangers and for the rail materials, but is not needed for the changes to the manufacturing line. Therefore, the procurement needs are considered low for this project. These assessments are now plotted on the bull's-eye chart shown in Figure 6.20. Next the following three steps are followed to tailor the base model.

First Step: The major milestones are laid out, and the project planning phase is defined using these steps: identify needs, define product specifications, and plan development tasks.

Second Step: The first criterion indicates which basic design model is used. For this project, the process model shown in Figure 6.11 is used, categorized as minor changes to an existing line.

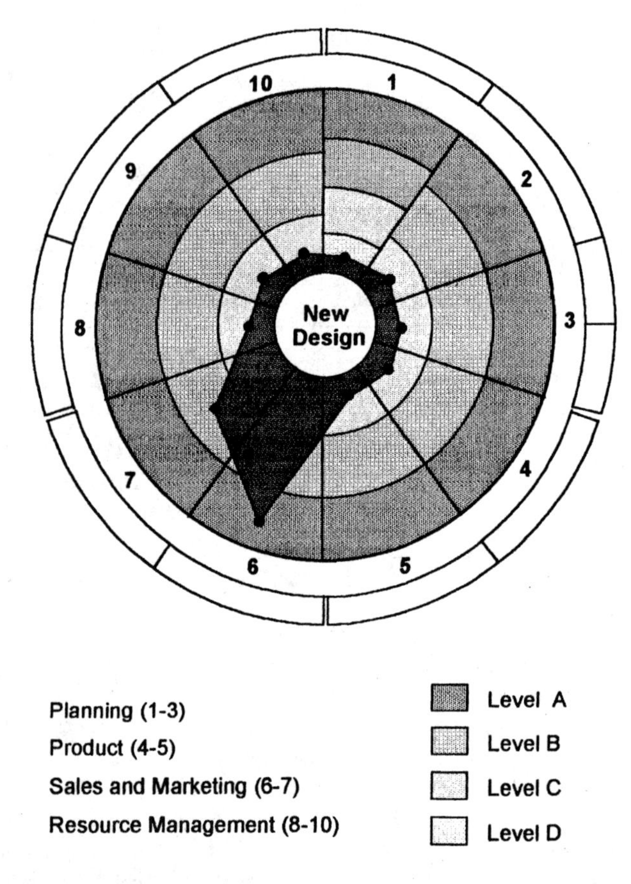

Figure 6.20 Process development target for the example [Adapted with permission from J. Phair].

Third Step: The remaining criteria indicate what changes need to be made to this base model.

Section 2: Manufacturing Process Complexity

No changes to the model are required.

Section 3: Standards and Regulations

No effect on the model.

Section 4: Product Design Paradigm

No effect on the model.

Section 5: Product Design Complexity

No changes to the model are required.

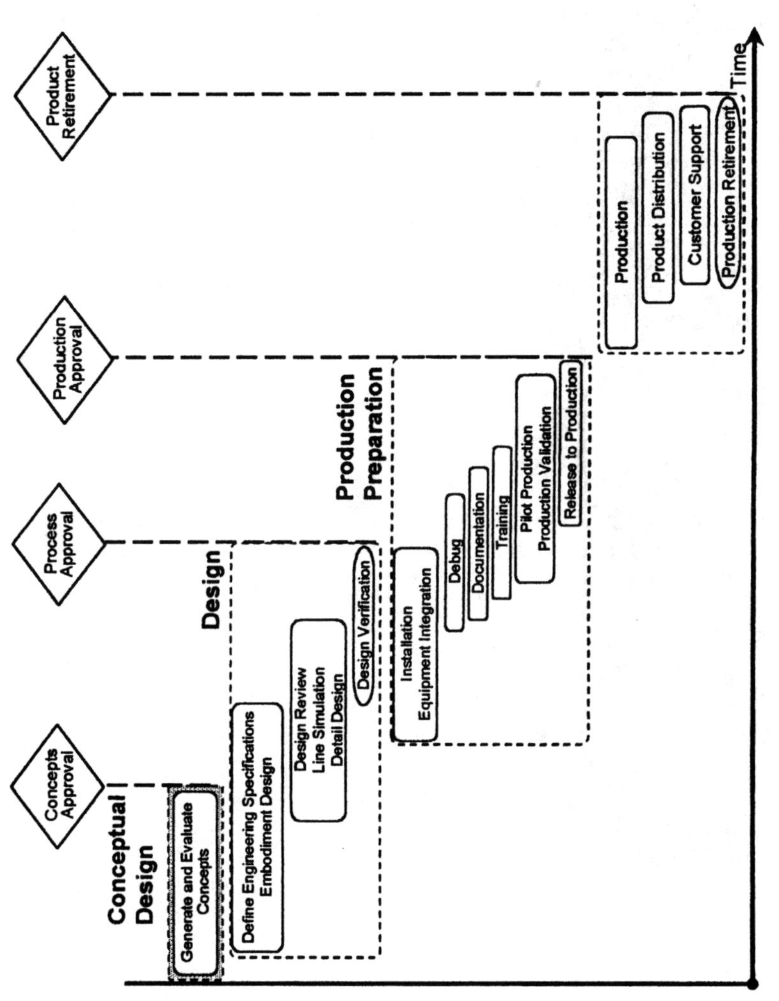

Figure 6.21 The tailored development model for the example [Adapted with permission from J. Phair].

Section 6: Time-to-Market
The delivery date is set by marketing in conjunction with the CDT and CPDT, with approval by the management team. From this date, the critical tasks are scheduled backwards, and more personnel are added to ensure that this date can be met.
Section 7: Sales Volume
There is no effect on the model.
Section 8: Supplier Involvement
No change to model.
Section 9: Partner Involvement
No effect on the model.
Section 10: Procurement Needs
No effect on the model or if the vendor selection and procurement steps are in the model, they may be removed.

The only modifications that need to be considered to the model are setting the dates of the milestones to meet the delivery target date. Otherwise, the tailored model in this case is the same as the base model. The tailored model is shown in Figure 6.21.

6.3 SUMMARY

The tailoring method presented here is unique. No other concurrent engineering methodology has proposed such a process. This tailoring allows the company to streamline the design process for both the product and the manufacturing process, which includes manufacturing, testing and packaging line development.

The first step to tailoring the product and process design methodology is to assess the new project using the criteria developed that asks questions about the product design, the manufacturing processes, sales and marketing information, and resources management. Once these questions are answered, the rated answers are plotted on bull's-eye graphs to indicate the amount of work required to attain the goal of having a new design or line. Next, a base development model is selected based on the project characteristics and that model is tailored based on the assessment. This tailoring process is intended to remove unnecessary design steps from the models, while still maintaining the integrity of the design information and history.

REFERENCES

Henson, C., *Defining the Outputs of a Small Business Concurrent Engineering Model*, Senior Thesis, University of Virginia, 1998.
Kemser, H., *Concurrent Engineering Applied to Product Development in Small Companies*, Masters Thesis, University of Virginia, 1997.

Phair, J., *Integrating Manufacturing and Process Design into a Concurrent Engineering Model*, Masters Thesis, University of Virginia, 1999.
Ter-Minassian, Natasha, *Concurrent Engineering in Small to Medium-Sized Organizations*, Masters Thesis, University of Virginia, 1995.

7

Assessing Risk in Design: An Example of Customization

In collaboration with Melanie Born

The complete concurrent engineering development methodology was described in Chapters 3-6. To make concurrent engineering work most effectively for your company, you may want to customize certain aspects of the methodology. In this chapter, a method for assessing the risk of the design of sheet metal and plastic parts is described along with how it fits into the CE methodology. This chapter will show how to incorporate industry-specific tools into the methodology. Chapter 8 presents additional tools within the methodology that allow the incorporation of the philosophy of industrial ecology into the development process through the use of design for environment tools. These two chapters to illustrate the flexibility of the methodology and how to customize the basic framework to fit your company's specific needs.

There is little work that addresses the assessment of risk within the product development process. It is generally assumed that if a systematic methodology is followed that risk is automatically minimized. This is true to some degree, however, risk assessment in product development is important enough that it should be addressed explicitly within the product development process itself. Although some issues of risk can be assessed generally, in most cases, the risk assessment of a design is material specific. There are several characteristics that a good risk assessment scheme should have. It should:

- Be general enough to be applied across different types of designs, but specific enough to give a good indication of problem areas within the design,
- Work within the CE development methodology framework,
- Be able to handle vague information at the beginning of the design process, and handle more specific issues later,

- Be organized in a logical manner, i.e., address risk issues related to manufacturing separately from design or resource allocation, and
- Be as easy as possible to use and to evaluate results.

The risk assessment model discussed in this chapter was developed in conjunction with a small group of designers that operate independently from the large corporation of whom they are a part. This group is responsible for the design of components for laser jet printers and other printer-related products. This group addresses product issues that include:

- Design,
- Prototyping,
- Testing,
- Selection and monitoring of suppliers responsible for building tools as well as those that perform contract manufacturing,
- Coordination of parts shipments from suppliers,
- Quality of manufactured components, and
- Financial issues including component pricing and tooling budgets.

The designers in this group, although aware of the risks in product and component development, did not use a formal approach to assessment. Rather, they used an *ad hoc* approach that was developed from personal experience. Using their experience, our own, and relying on published literature, the following methodology was developed and then tested. In this chapter, we will present the methodology in the first half, then present two design examples in the latter half to show how the methodology can be used.

7.1 THE RISK ASSESSMENT MODEL FOR SHEETMETAL AND PLASTIC PARTS

The model that is presented in this section is in the form of a checklist and is in an abbreviated form for clarity of presentation. See Appendix B for the complete risk assessment model. The model, as shown in Figure 7.1, is divided into four sections: general questions, level one questions, level two questions, and the risk assessment graphs. In each succeeding level, the questions become more detailed about the design, manufacture, and tooling for the part. The risk assessment graphs allow the team to review the information graphically and decide whether the design needs to be modified by iterating back to an earlier design step. Figure 7.2 shows how this model fits into the CE product development model. In this figure, we see that the general questions and the level one questions are answered within the conceptual design phase in the generate concepts design step. The general questions will be handled at the start of the generate concepts step, and the level one questions will be answered at the end of the step. The effects of the risk

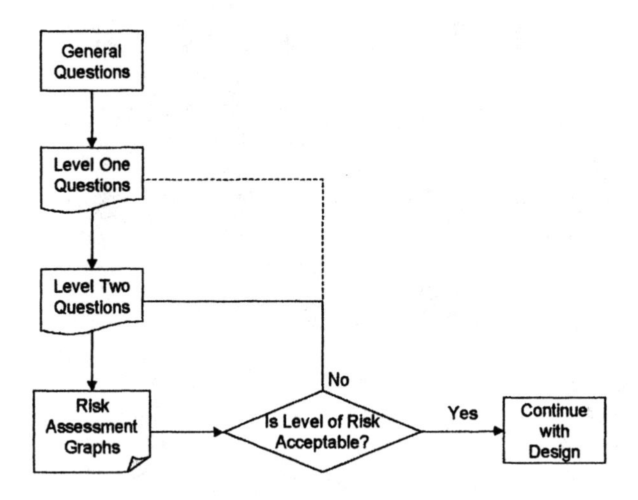

Figure 7.1 Overview of the risk assessment model [Reprinted with permission from M. Born].

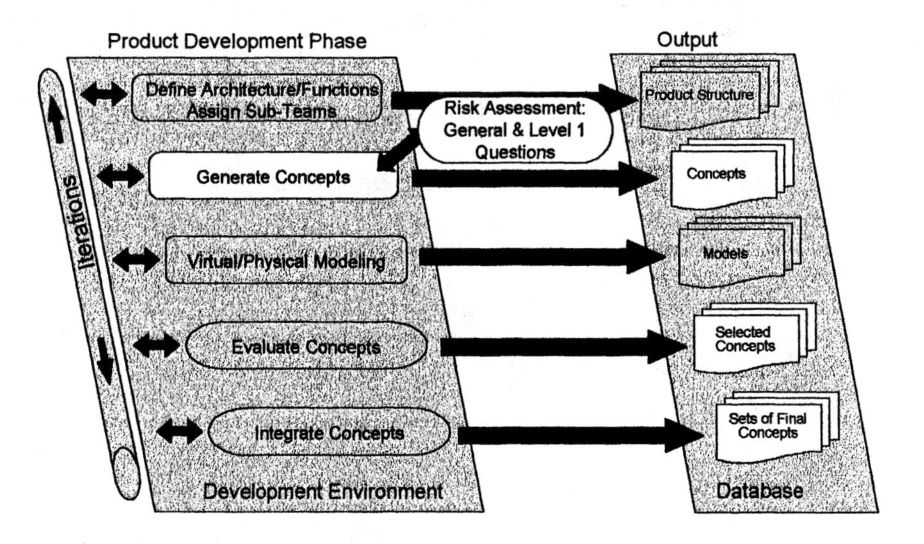

Figure 7.2 The product development model's conceptual design phase, showing inclusion of the risk assessment model.

assessment on the design is shown by the highlighted white output steps: concepts, selected concepts, and final concepts. The level two questions, which are much more design specific, are associated with the design phase, shown in Figure 7.3. In this figure, the level two questions are associated with embodiment design. The risk assessment graph can be evaluated both at the end of embodiment design and again in the design review. If the team decides that the level of risk is too high for some aspects of the design, they can revisit embodiment design, changing those aspects that reduce the risk of having problems with the design later in the product's life. This iteration process can continue until the level of risk has been reduced to an acceptable level. However, the team must also remember that the part design must also meet the scheduling constraints of the design process as a whole. Therefore, iteration cannot continue indefinitely. There is a trade-off between meeting schedule deadlines and reducing risk. In addition, there is also a chance that a change in the design of one part may affect the design of other

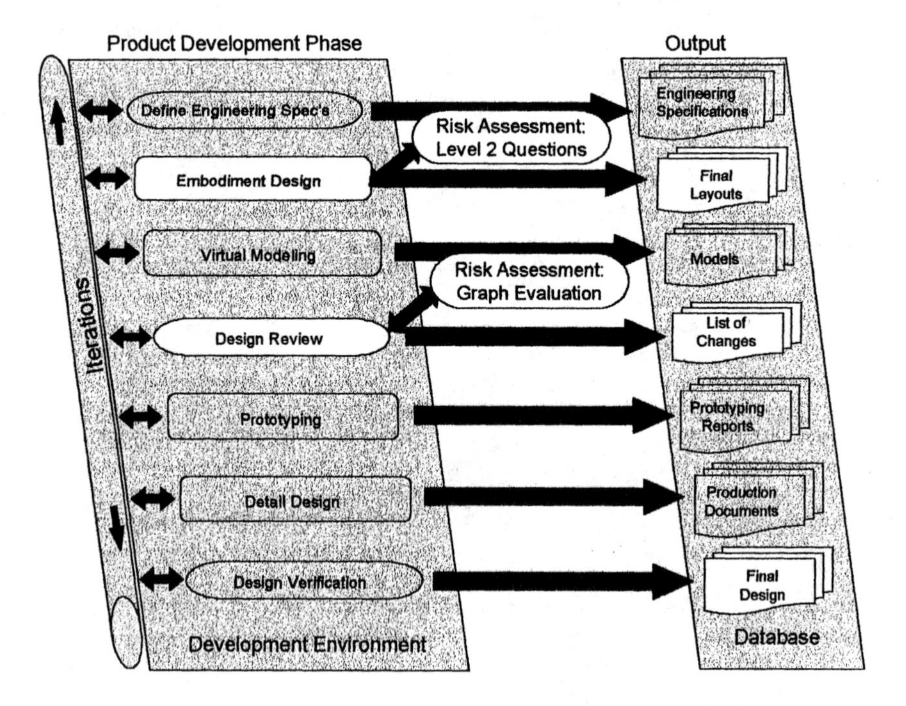

Figure 7.3 The product development model's design phase, showing inclusion of risk assessment model.

components, increasing another part's level of risk. The team must be aware of these trade-offs and have open lines of communication regarding component designs and their changes. Open communication is a key characteristic of good CE design practice. Finally, the design team should refrain from moving between levels. If the team revises answers from level one, this change could impact multiple component designs. As a result, the team will spend valuable design time addressing the impacts of this change, rather than spending their time moving forward with the design. For this reason, iteration is only encouraged to reduce risk in the parts indicated by the risk assessment graphs, the concept of which is shown in Figure 7.1 by the lines returning to level two and level one. Neither of these lines have arrows, indicating that the team should show caution in iterating through the questions and making major changes, especially if they decide to return to level one. Next, each of the model sections will be described in more detail.

7.1.1 General Questions

The general questions, shown in Table 7.1, ask about the component's expected development time frame, the expected due date for delivery of a functional component, and the estimated cost. In Table 7.1, the question to be answered is in the left column. A space for the answer is in the middle column under the time/cost heading, and the right column allows for comments for justification of the answers. The answers provided in this section should guide the team through the design process. The team needs to avoid making decisions in part design that will lead to inconsistencies with these answers. This table should become part of the output for this design step. A modified output for this design step will be shown in the next section.

Table 7.1 The general questions section [Reprinted with permission from M. Born].

General Questions		
Questions	**Time/Cost**	**Comments**
1. What is the expected time frame for the development of this component?		
2. When will the functional component be ready for testing?		
3. What is the estimated cost of the component?		

7.1.2 Level One Questions

In this section, there are two types of questions to be answered: informational questions and risk assessment questions. The informational questions, shown in the white sections in Table 7.2, are asked to encourage the development team to think about product specific characteristics before assessing the level of risk in the component or product. The risk assessment questions, shown in the light gray sections, ask the team to make a qualitative judgement about the perceived risk of different aspects of the design and the design process. The questions in this section are grouped into three sub-sections: design, manufacturing, and tooling. The design questions are answered as part of the generate concepts design step in the design model. The manufacturing and tooling questions are to be answered as part of the generate concepts design step in the manufacturing process development model (see Figure 5.5). These questions will be answered by the cross-functional process development team with input from the product development team.

As in the general questions section, the question is shown in the left column of Table 7.2. For the some of the questions, the options for the answers are shown in the options column. The team should answer the questions in the answers column and add any justification or explanation needed in the comments column. For some of the informational questions, as with question 5, if the team answers 'yes', then they should proceed to question number 6, skipping 5a, 5b, and 5c. For some of the risk assessment questions, the options column contains a set of answers from which the team chooses. Each answer has a corresponding point value ranging from 0 to 4, in which 0 represents a low risk, and 4 represents a very high risk. The team should place the value of the chosen answer in the answers column. It is important that the team use the number scheme given for the answers in these risk assessment questions, as these values will be used later in the risk assessment graphs.

For the manufacturing and tooling set of questions, the team is instructed to rate possible suppliers in a manufacturing supplier matrix. (See questions 13 and 14 in the manufacturing section, and questions 17 and 18 in the tooling section in Appendix B). The supplier matrix, shown in Table 7.3, suppliers are rated using the following categories:

- Necessary personnel needed to provide components or equipment,
- Available equipment for supplying components or tools,
- Availability of necessary technology,
- Quality monitoring capabilities,
- Financial stability ,
- Material sourcing capabilities, and
- Post delivery support in the case of the tooling supplier matrix.

These matrices should help the team compare suppliers, thereby making the selection more straightforward.

Table 7.2 Level one questions section [Reprinted with permission from M. Born].

	Level One Questions			
	Questions	**Options**	**Answers**	**Comments**
D	1. Describe what the function this component will perform.			
E	2. How well do you understand the function you just described?	a. Very well (0 pts) b. Well (1 pt) c. Somewhat well (2 pts) d. Undecided (3 pts) e. Do not understand (4 pts)		
S	3. What type of design is it?	a. Incremental (0 pts) b. Evolutionary (1 pt) c. Original (3 pts) d. Breakthrough (4 pts)		
I	4. Should the part be one piece or an assembly?			
G	5. Will this product be standalone?	a. Yes b. No Note: If yes, then go to question #6.		
N	5a. Where within the final product is the part located?			
	5b. Explain how it interacts with mating parts.			
D	5c. With these interactions in mind, how stable do you expect the system to be?	a. Very stable (0 pts) b. Stable (1 pt) c. Somewhat stable (2 pts) d. Not sure (3 pts) e. Not stable (4 pts)		
E	6. How durable are the parts expected to be? (This will affect material selection)			
S	7. Will you use sheetmetal or plastic?			

Table 7.3 Manufacturing supplier matrix [Reprinted with permission from M. Born].

	Tooling Supplier Matrix							
	Name	Necessary Personnel	Available Equipment	Necessary Technology	Quality Monitoring (Cp, Cpk)	Financial Stability	Material Sourcing	Post-Delivery Support
1.								
2.								
3.								
4.								
5.								

The product development team is encouraged to perform as many iterations as needed within the level one questions section until all members of the team feel they have addressed the high risk items as best they can at this stage of the design process. Once finished, the team can proceed with the development process by completing the concept generation step and completing its associated output. The output for this design step is shown in Table 7.4.

Table 7.4 Modified output for the design step: generate concepts.

Title	Product concepts
Definers: Primary	Cross-functional product development team
Definers: Support	New technology team, suppliers
Overview	Sketches and descriptions of selected concepts General questions table Level one questions table
Documentation	Sketches of generated solutions and concepts
	Concept evaluation matrix

7.1.3 Level Two Questions

In the level one questions, the product development team chooses the component's material to be used as well as its associated manufacturing process. Because of this, the level two questions are divided into two categories, those related to sheetmetal parts and those for plastic parts. As in the level one questions, the level two questions are divided into the categories of design, manufacturing, and tooling. Again, the layout of the questions is the same as in the level one questions as seen in Table 7.5, which shows a few questions from the sheetmetal questions in the design category. The same four columns are present: questions, options, answers, and comments. The same process for answering the questions is followed as for the level one questions.

7.1.4 Risk Assessment Graphs

The risk assessment graphs serve as a visual representation of the answers supplied in the questions sections. With a visual representation, the development teams can quickly identify the areas of risk within the product, its manufacture, and tooling. Figure 7.4 shows the types of graphs contained in this section of the risk assessment method. The graphs are divided into the groups we saw in the level two questions, sheetmetal and plastics. Both these groups are then divided into the level one questions section and the level two questions section. The level two question graphs are then further sub-divided into design, manufacturing, and tooling.

Figure 7.5 shows a graph of design risk assessment questions for a fictional sheetmetal part. In the shaded half of the graph, the fastener options are rated. From the figure we see that the rivet has the lowest score among the options for functionality/performance, whereas the extrude-and-roll and the tab options have the highest scores, while Tog-L-Loc and Spot weld scored in the middle range. From this information, the team can decide upon the type of fastener that best fits the product design. With several methods from which to choose, the team can perform a trade-off analysis to look for the best fit. For example, they may decide anything scoring higher than a 2 is unacceptable, therefore, they can then choose between Tog-L-Locs, rivets, or spot welds as fastening options. If the cost of the fastener is a critical issue, then the team must balance the risk of the fastener with its cost. In addition to the fastener options, several other issues are addressed in this graph, as seen in the white section. First, the part that is rated a 2 for its compatibility with the assembly method signals the team members that either the assembly method should be modified to better accommodate the part design, or the design should be changed to be more compatible with the assembly method. If after examining both options, the team finds either of the changes leads to an increased risk in other categories, or the costs are too high, they can then decide to leave the part design unchanged. Another category that should be assessed is the ease of disassembly. Again, the team should look at the part design and determine if this category is important to the design, in other words, is it destined to be disassembled, and if so, can the risk be lessened through a design change?

It is important that if any changes are made to the part design that these changes be communicated to other team members and other development teams. In addition, the team should return to the questions section corresponding to the graph and to ensure any changes required in the questions section are made.

Table 7.5 Level two questions design section for sheetmetal parts [Reprinted with permission from M. Born].

	Level Two Questions: Sheetmetal			
	Questions	**Options**	**Answers**	**Comments**
D	1. Determine the shape and dimensions of the part.			
E	2. Identify the critical part dimensions.			
S	3. Does this part require special insulation properties? If so, describe them.			
I	4. What type of sheetmetal should be used? Should it be corrosive or erosive?			
G	5. What are the cosmetic requirements for the part?			
N	6. Given these cosmetic requirements, should a pre-plated material or a post-plated material be used?			
	7. If you need to join sheetmetal parts or corners, rate the following aspects in terms of needed functionality/performance.	a. Very effective (0 pts) b. Effective (1 pt) c. Somewhat effective (2 pts) d. Not sure (3 pts) e. Not effective (4 pts)		
D	7a. Tog-L-Loc			
E	7b. Extrude and roll			
S	7c. Rivet			
I	7d. Spot weld			
G	7e. Tabs			
N	8. Considering the chosen material, the design, and the process, how confident are you that your durability expectation will be met?	a. Very confident (0 pts) b. Confident (1 pt) c. Somewhat confident (2 pts) d. Not sure (3 pts) e. Not confident (4 pts)		

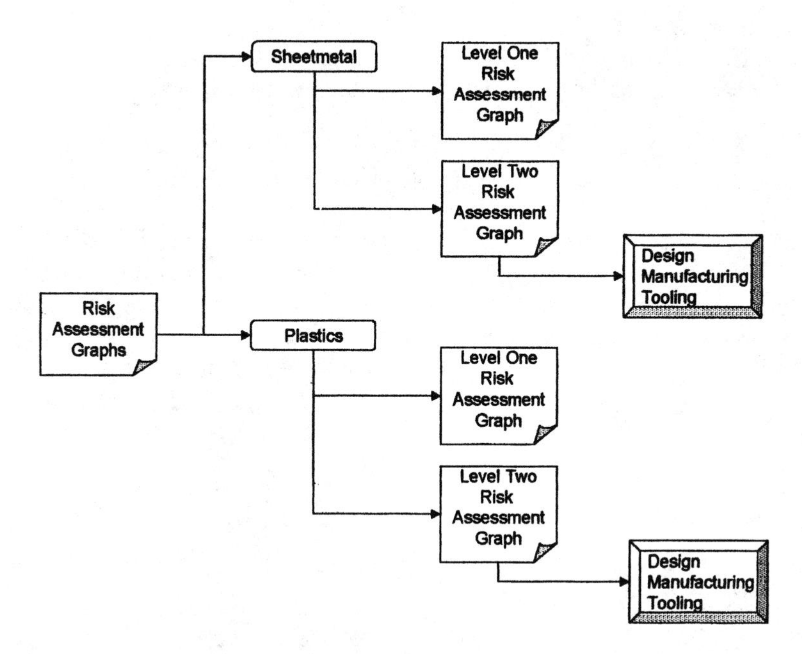

Figure 7.4 Organization of risk assessment graphs [Reprinted with permission from M. Born].

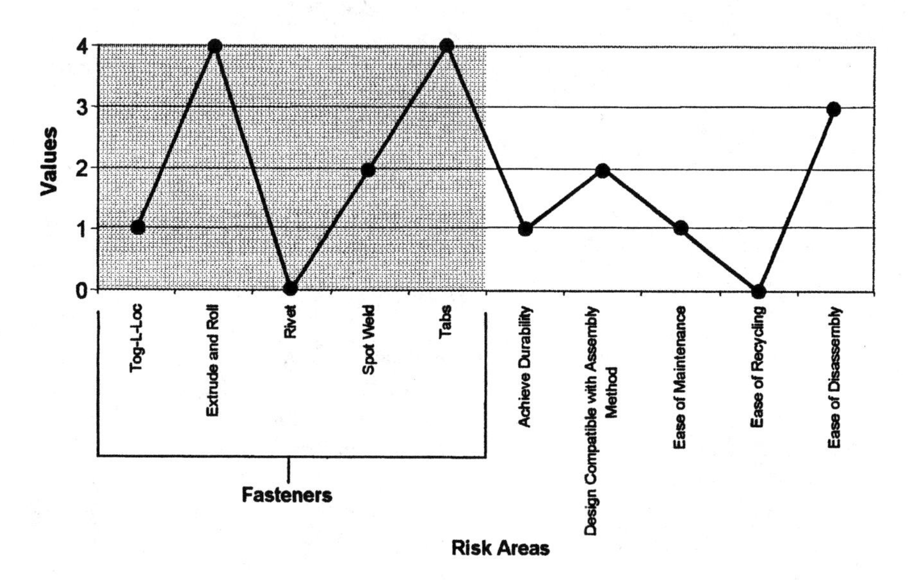

Figure 7.5 Risk assessment graph of level two questions for a sheetmetal part [Reprinted with permission from M. Born].

Finally, if any changes are required, then the team needs to examine their impact on other product parts, and look at any new problems that might arise out of these new changes.

There are two remaining steps in this section. One is that a lessons learned section is included in which the team is asked to document the changes made and what they learned through this process about the part and its associated risks. This documentation should become part of the output for the design step, design review, as shown in Table 7.6. This output is an important document for reference by future design teams. The documentation can provide a clear picture of the decisions made, their impact on the design, and once the product is in production, an assessment made of how the decisions impacted the final design can be ascertained.

Table 7.6 Output for the design step: design review.

Title	List of changes
Definers: Primary	Cross-functional product development team
Definers: Support	Cross-functional process development team, customer, field service
Overview	List of changes to be made to meet specifications and to accommodate design integration, manufacture, test, packaging, and field service
Documentation	Risk assessment graphs
	Evaluation results and rationale behind suggested changes
	Parts and documents affected by changes
	Lessons learned

7.2 EXAMPLES OF THE USE OF THE RISK ASSESSMENT MODEL

In the following sections, the risk assessment model is illustrated with the use of two examples. The first example is a sheetmetal part used to house electronic components, and the second is a plastic part that resides on a circuit board. These examples should give the reader a better picture of how to use the risk assessment model and how the model can be used to reduce the risk of a particular component. These examples will concentrate on the use of the risk assessment model, and will not illustrate the use of the CE methodology, per se.

7.2.1 Design of a Sheetmetal Component

An electronics company needs to design a sheetmetal component for housing electronic components for one of their products. The company's goal is to design the component, build the necessary tooling to manufacture the part, test the component, and start manufacturing operations within 16 weeks. The product for which this part is destined is to be distributed both in the US and in Europe, and the company has made the decision to use both a US and European supplier for the manufacture of the part.

The first step is for the product development team to start with the general questions section. The answers are shown in Table 7.7. Next, the team answers the questions from the level one questions section shown in Table 7.8. After answering the level one questions, the team then analyzes the risk assessment graph that corresponds to these questions. Analyzing this graph, shown in Figure 7.6, the team finds that the highest risk area is product stability, which scores a one. Product stability is defined as the interaction between the component that is under development and the product into which it will be installed. Because this score is so low, the team decides to proceed with the design and to move on to the level two questions.

After answering the level two questions, the product development team analyzes the risk assessment graphs corresponding to these questions. These graphs are shown in Figure 7.7-7.9, and correspond to the sub-sections: design, manufacturing, and tooling.

In the risk assessment graph for the component design, shown in Figure 7.7, the fastener options that scored the lowest for joining the sheetmetal parts are the spot weld and the tabs. Therefore, the team decides to choose one of these methods.

Table 7.7 The general questions for the sheetmetal example [Reprinted with permission from M. Born].

General Questions		
Questions	**Time/Cost**	**Comments**
1. What is the expected time frame for the development of this component?	16 weeks	This time was determined by the Sales and Marketing group.
2. When will the functional component be ready for testing?	10 weeks	The team feels this time frame is reasonable given past design experience.
3. What is the estimated cost of the component?	$0.30	Based on other company part costs.

Table 7.8 Level one questions section for the sheetmetal part [Reprinted with permission from M. Born].

	Level One Questions			
	Questions	Options	Answers	Comments
D	1. Describe what function this component will perform.		Case to enclose electronics	
E	2. How well do you understand the function you just described?	f. Very well (0 pts) g. Well (1 pt) h. Somewhat well (2 pts) i. Undecided (3 pts) j. Do not understand (4 pts)	0	
S	3. What type of design is it?	e. Incremental (0 pts) f. Evolutionary (1 pt) g. Original (3 pts) h. Breakthrough (4 pts)	0	
I	4. Should the part be one piece or an assembly?		One piece	
G	5. Will this product be standalone?	c. Yes d. No Note: If yes, then go to question #6.	No	
N	5a. Where within the final product is the part located?		Inside the machine	
	5b. Explain how it interacts with mating parts.		Case slides into machine	Grooves on side of case for anchor
D	5c. With these interactions in mind, how stable do you expect the system to be?	f. Very stable (0 pts) g. Stable (1 pt) h. Somewhat stable (2 pts) i. Not sure (3 pts) j. Not stable (4 pts)	1	
E	6. How durable are the parts expected to be? (will affect material selection)		Case should last 10 years	This is expected life of machine
S	7. Will you use sheetmetal or plastic?		Sheetmetal	

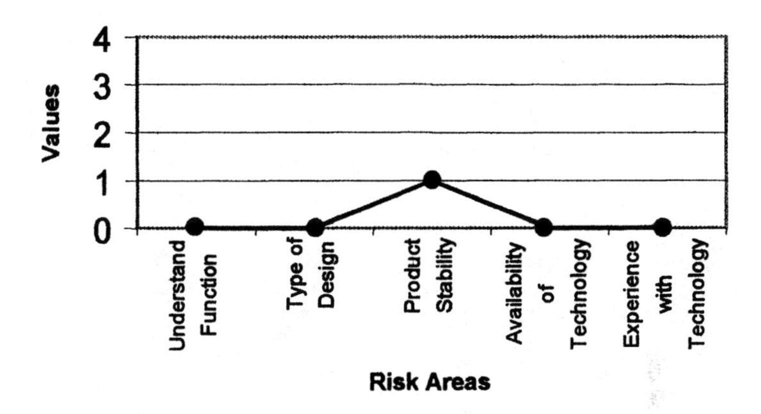

Figure 7.6 Risk assessment graph of level one questions for the sheetmetal part [Reprinted with permission from M. Born].

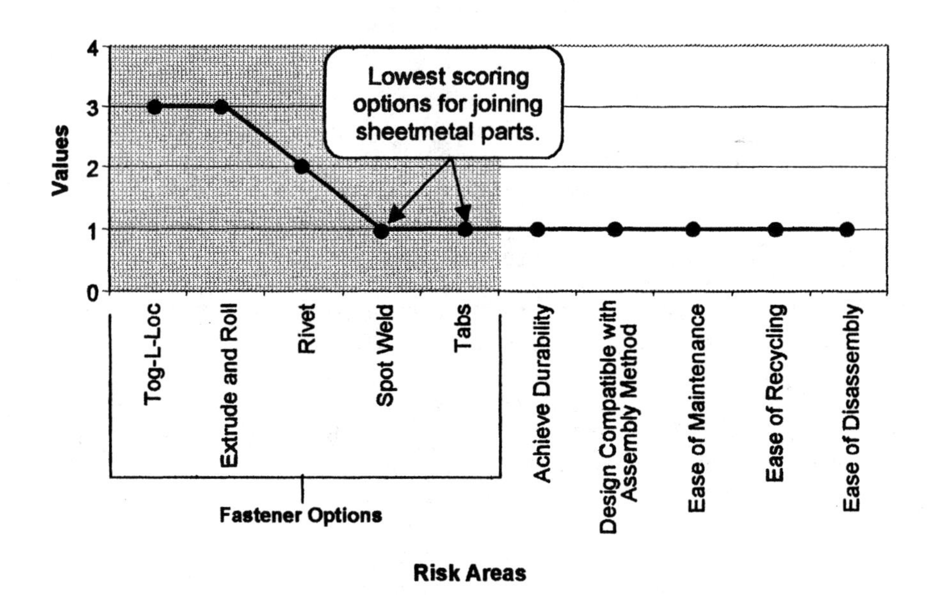

Figure 7.7 Risk assessment graph of level two questions for the sheetmetal part design [Reprinted with permission from M. Born].

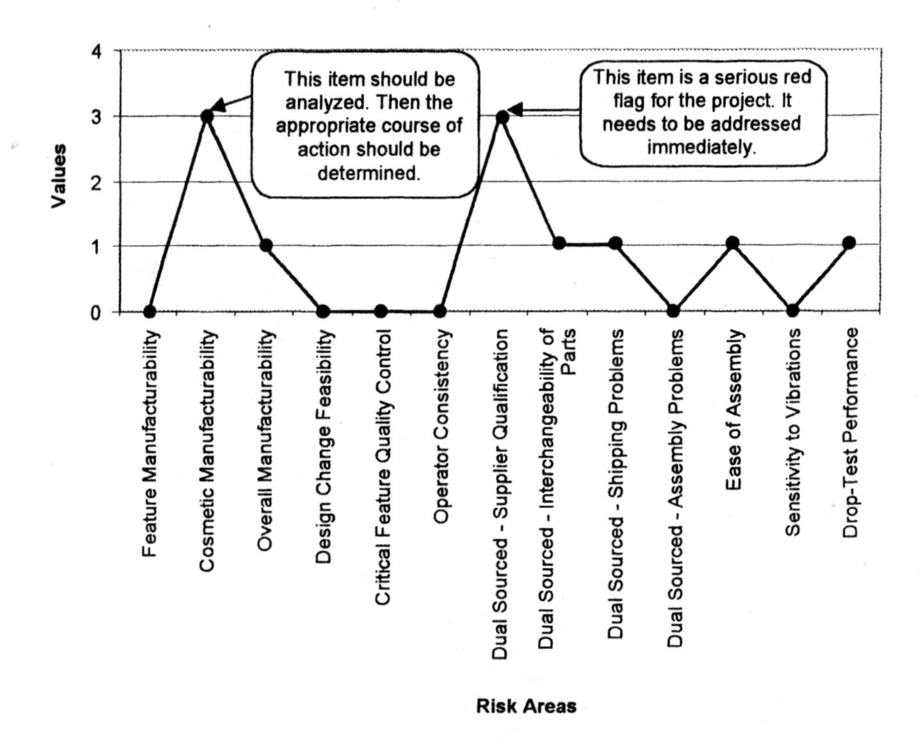

Figure 7.8 Risk assessment graph of level two questions for the manufacture of the sheetmetal part [Reprinted with permission from M. Born].

In the risk assessment graph for manufacturing the housing, shown in Figure 7.8, there are two areas of concern: cosmetic manufacturability and supplier qualification for dual sourcing. For the first item, the development team is unsure if the cosmetic specifications can be achieved. However, since this housing part will be inside the machine where it will not be visible to the customer, the team needs to decide if the level of surface finish that can be achieved is acceptable. The team may need to revisit the product's engineering specifications to ensure that the surface finish specifications are not too tight for the application, which is a part that is not readily visible to the customer. If after review, the specifications must remain as they were written, then the team must determine if varying some aspect of the manufacturing process or the material can result in the desired surface finish.

In the second item, the team uncovers a qualification problem with the suppliers of the part. Both suppliers must have the capability of delivering a high-quality, finished component. Due to the slight differences in manufacturing practices in the US and Europe, it may be easier to qualify one supplier over the other. If one supplier can be approved, then the team may want to consider getting these vendors to cooperate through a partnership agreement for the manufacture of this part. If this cooperation is not possible or agreeable to the companies, then the other two options are to waive some specifications for the vendor that is not in compliance, or the team can search for an alternate vendor. All these actions should be considered as a means to reduce the risk.

Figure 7.9 shows the risk assessment graph of the level two questions for the tooling sub-section. In this graph, a red flag is raised with volume compliance. The process development team is not sure that the needed volumes can be achieved with the available production tools. Meeting needed volumes is a critical issue for the company, and a solution for this problem must be found. There are several options open to resolve this problem. One, which is the most costly, is that the team can consider trying to justify the purchase of another production tool to satisfy the volume requirements. Two, they may be able to address the shortfall in production by scheduling changes such as adding a shift, if that is possible. If the first two options are infeasible, then the team can consider using the prototype tool to meet productions. While this is the least attractive of the options, it may be the most cost effective. It may be possible to modify the prototype tool to withstand the wear and tear of production by coating the tool. However, if it is not made for high volumes, this may be a short term, back-up solution only. In any case, the team must search for a solution to this problem in conjunction with the management team.

With this short example, we can see that the development teams can help

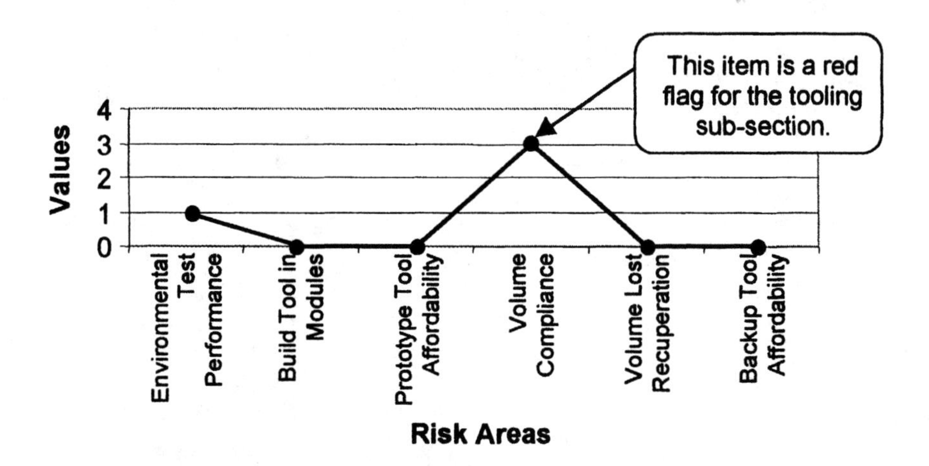

Figure 7.9 Risk assessment graph of level two questions for the tooling for the sheet-metal part [Reprinted with permission from M. Born].

systematically identify risks that could jeopardize the success of a product launch. The graphic outputs are especially useful for a quick visual reference to the long tables that must be filled out (shown in Appendix B).

7.2.2 Design of a Plastic Component

The same company that designed the housing also needs to design a part that will separate two components on a circuit board, which will reside in the housing that was discussed in the last section. All the circuit boards will be manufactured domestically. This is a simple part design that should be developed within 10 weeks, and ready for testing in six. As in the last example, the development team will begin the risk assessment for this part during the conceptual design phase, beginning with general questions in the generate concepts design step. These questions are shown in Table 7.9.

Next, the team proceeds to the level one questions, and its assessment graph, which is shown in Figure 7.10. As we see from this figure, the highest scoring item is the type of design, because the development team identified this component design as being new to the company. To better understand the requirements for this component, the team would be advised to benchmark similar designs from other companies. Otherwise, the graph indicates this is a low risk part.

After resolving the issues associated with the level one questions, the team can proceed with the design process. Once the team has reached the design phase, as part of the embodiment design step, the team should answer the level two questions. The risk assessment graphs for these questions are shown in Figures 7.11-7.13. Figure 7.11 shows the graph for the plastic part design. In this graph we see that the highest score is associated with ease of routine maintenance for this part. The reason this category scores high is that the part is not easily accessible. Ordinarily, this score would indicate a problem, however, since this component

Table 7.9 The general questions for the plastic component example [Reprinted with permission from M. Born].

General Questions		
Questions	**Time/Cost**	**Comments**
1. What is the expected time frame for the development of this component?	10 weeks	This time was determined by the Sales and Marketing group.
2. When will the functional component be ready for testing?	6 weeks	The team feels this time frame is reasonable given past design experience.
3. What is the estimated cost of the component?	$0.02	Based on other company part costs.

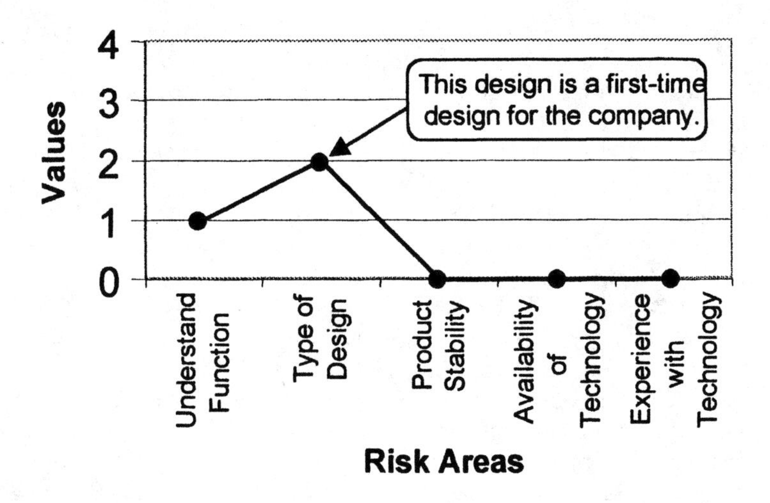

Figure 7.10 Risk assessment graph of level one questions for plastic part [Reprinted with permission from M. Born].

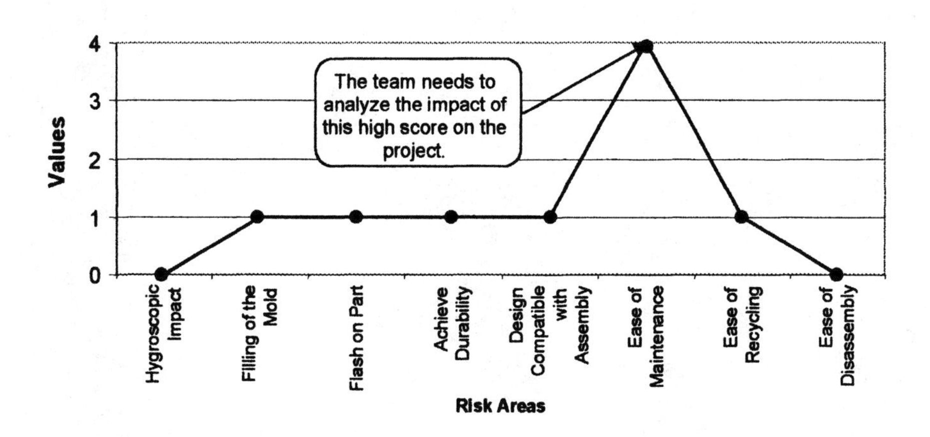

Figure 7.11 Risk assessment graph of level two questions for the design of the plastic part [Reprinted with permission from M. Born].

resides on a circuit board, it is not meant to undergo routine maintenance. If this component breaks, a service call would be needed, as with any other part on the circuit board.

Figure 7.12 shows the risk assessment graph for the manufacture of the plastics part. There is only one area of concern that we see on this graph: environmental test performance. This is a suite of tests that indicates how the part will perform under certain operating conditions, such as hot and dry, hot and wet, cold and dry, and cold and wet. In some of these test conditions, the part was found to warp, which is of some concern. The team must now determine how this warpage will affect the performance of the part, and what changes can be made to lessen the effects. They may be required to change materials or part shape to decrease or eliminate this problem.

Figure 7.13 shows the risk assessment graph for the tooling for the plastic component. In this graph we see that the main problem arises from the inability to meet the volumes required. As we saw in the previous example on the sheetmetal part, there are several ways to handle this issue. The team can invest in more production machines; however, this is an expensive proposition, and must be cost-justified. This justification may be difficult considering the cost of this part. Two, the team can consider contracting the manufacturing to a supplier, which again will raise the cost of the part. Third, the team can look at changing the production schedule to add a shift either during the week, or on a weekend if needed. As in the last example, the development team must discuss these issues with the management team, and recommend a solution.

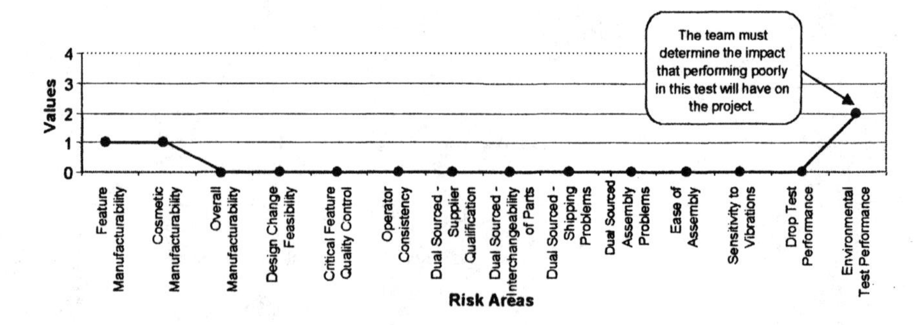

Figure 7.12 Risk assessment graph of level two questions for the manufacture of the plastic part [Reprinted with permission from M. Born].

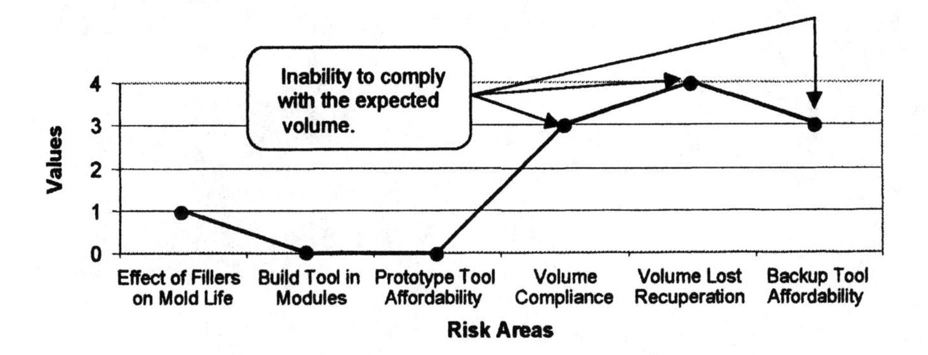

Figure 7.13 Risk assessment graph of level two questions for the tooling for the plastic part [Reprinted with permission from M. Born].

7.3 SUMMARY

The risk assessment model presented in this chapter uses a series of questions that are integrated into the design process. This model is an example of how a company can integrate their design procedures to fit within the concurrent engineering methodology. As we saw at the beginning of this chapter, the general questions section and the level one questions section were integrated in the conceptual design phase of the CE methodology in the generate concepts design step. The more detailed questions regarding the design were placed in the design step, embodiment design, in the design phase. The tooling and manufacturing questions were integrated into the embodiment design step in the manufacturing process model. In all these cases, the output for these steps had to be modified to incorporate new information, so that the documentation would become part of the data to be shared between teams and among team members, and would also be available for future development teams.

This risk assessment model is a useful tool that can pinpoint areas of high risk in the development of new parts and processes for a product. Using the risk assessment graphs, the team can easily spot those areas of high risk and know what items need to be revisited to improve the success of the product. It is not an all-inclusive model, but can be added to and modified to fit a company's needs, as is the case with the CE methodology in general.

REFERENCE

Born, M., The Assessment of Risk in the Development of Sheetmetal and Plastic Parts, Masters Thesis, University of Virginia, 2000.

8

Integrating Industrial Ecology into the Concurrent Engineering Framework

In collaboration with Mark Sondeen

Companies have been pressed since the 1960s to reduce their emissions and waste output from their plants. They must abide by EPA regulations, as well as regulations from a host of other government organizations such as OSHA and NIOSH. The traditional approach within a company is to regulate the output of harmful materials, however, in the late 1980s a new approach was introduced. Industrial ecology takes the view of the company as an eco-system [Frosch and Gallopoulos, 1989]. In this system, the usage of energy and materials should be optimized and waste generation minimized. Ideally, the waste streams from one process should serve as materials for another process, thereby allowing the company to obtain as much monetary value from its raw materials and processed materials as possible. How does a company begin this process? In general, design for the environment (DFE) proposes a set of tools and methodologies that can help guide a company to include environmental objectives into their purchasing, design, and manufacturing processes [Allenby, 1994].

The concept of industrial ecology may sound burdensome, however, there are many benefits to the company in undertaking this approach. The company can gain strategic advantage, not the least of which is financial, associated with a reduction of operating and production costs through waste handling and disposal, fulfilling or avoiding regulatory requirements, future liability, the acquisition of raw and outsourced materials, and the sale or disposal of by-products or waste. Furthermore, a company may gain new market share through superior products due to product characteristics of energy efficiency, upgradeability, durability, and fewer harmful byproducts or materials.

In this chapter, we will introduce a methodology that can be used in conjunction with the concurrent engineering methodology and that will allow a small company to incorporate DFE into the design of their products and processes. These tools are mainly used in the redesign of products and processes through the evolutionary design process, when new products are introduced based on previous product lines. The chapter begins with an overview of the DFE methodology and its relationship to concurrent engineering. Next, specific tools are introduced and associated with their appropriate phase in the CE methodology. Finally, an example will be given to show how the methodology can be used in an industrial setting.

8.1 METHODOLOGY OVERVIEW

We assume that a company will use this methodology to improve their environmental profile of current product lines and their associated manufacturing processes. Therefore, as with an evolutionary design of a new product, the design history of the product line is assumed to be known. Figure 8.1 shows an overview of the methodology, which is assumed to begin with the production/service phase of the CE methodology. Within this phase, the company can gain a clear understanding of the characteristics of the existing product and processes from an environmental standpoint. An inventory is performed in this phase of the CE process. In the project planning phase for a follow-on product, prioritization tools are used to rank the DFE tasks to be accomplished in this design cycle. In the conceptual design phase, tools are used to improve the product and process environmental performance. At the end of these phases, most of the DFE considerations

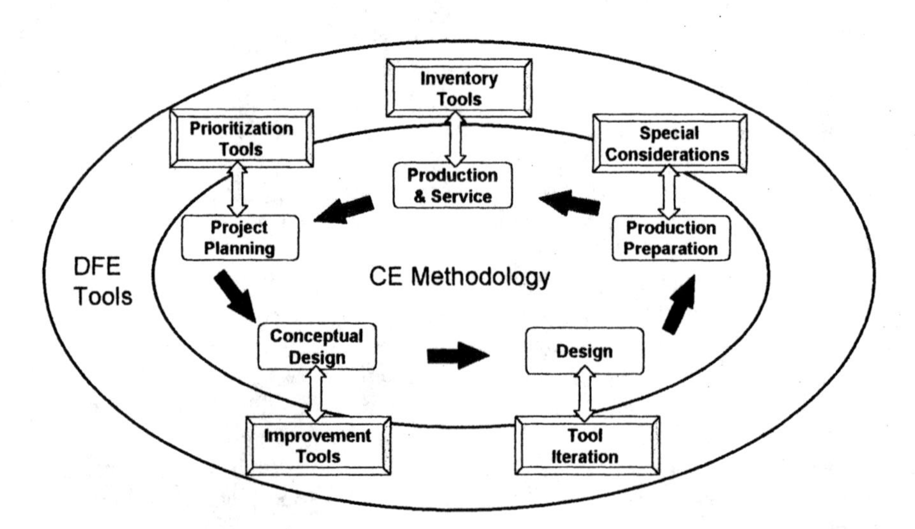

Figure 8.1 Methodology overview [Reprinted with permission from M. Sondeen].

have been made. The work in the other two phases will be the implementation of the product design changes and iteration to reconsider choices (if problems are found with implementation). Next, each of the design phases will be discussed in relationship with the associated DFE tools. Each tool will be identified, defined, and incorporated into the project activities described. Since this DFE methodology begins with the production and service phase, that is where this description will start. We will then follow the phases around the diagram as shown in Figure 8.1.

8.1.1 Production and Service Phase

As we have seen in this book, a new project typically begins with a kick-off meeting followed by the project planning phase. However, as we stated earlier, in this case, we begin with the production and service phase of the current product to gain an understanding of where the company currently stands from an environmental point of view. Both the CDT and the CPDT will work with the PT to understand the current product and its manufacturing processes, and these teams will then go on to design the follow-on product. Figure 8.2 shows the production and service phase with the associated DFE tools matched with the appropriate design steps. There are three tools associated with the production design step: the life cycle map and the sphere of influence, the process flow diagram, and material and energy balances. The voice of the customer is used with the customer support design step. There are two tools used with product retirement: end-of-life (EOL) fate, and the EOL strategy plot. Input from the PT as well as the outputs from the design steps will provide valuable information for this inventory work. Next, each of the tools will be discussed in more detail.

Life Cycle Map and the Sphere of Influence

An important aspect of industrial ecology is understanding a product's life cycle, that is, looking at all the life stages of the product from raw material acquisition to the product's end-of-life. Only by looking at all these stages can we begin to understand the product's impact on the environment in its totality. The sphere of influence is a DFE tool used to capture this picture for the company. First, we have defined sixteen stages to help understand the life cycle of a product. These stages are shown in Figure 8.3. Using these sixteen stages, the CDT, CPDT, and the PT should attempt to plot the life cycle of the product from acquisition of all raw materials through product end-of-life. This can be a complex undertaking, as typically one manufacturer is only one of many companies that is associated with the product and its parts and manufacture. Raw materials, purchased components, and packaging materials arrive from suppliers, and the end product may not go to the end user, but may go to another company to be used as a component in another product. For example, a company that produces disk drives sells their product to companies that then use them in their products, usually computers. As a result,

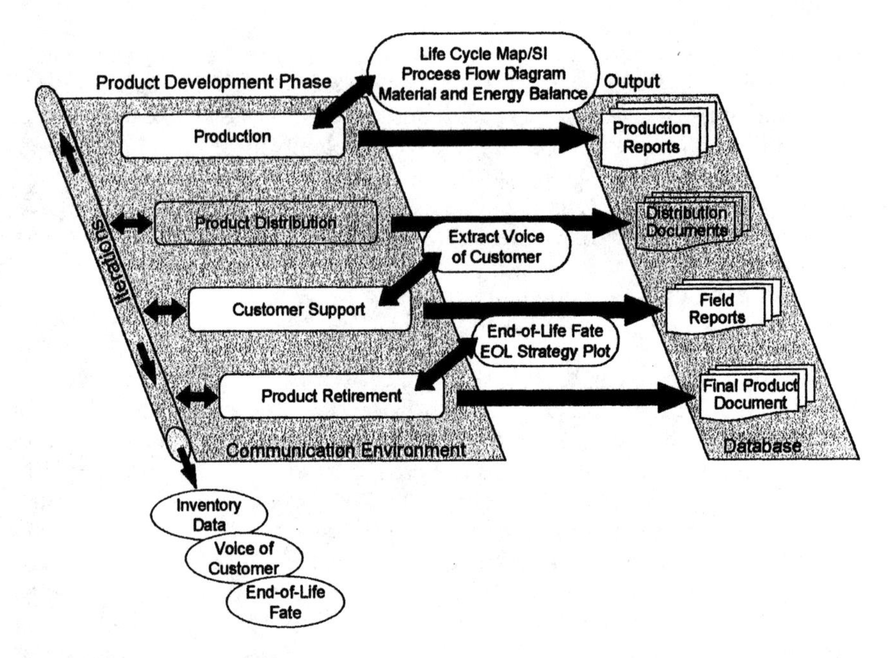

Figure 8.2 Production and service phase showing associated DFE tools [Reprinted with permission from M. Sondeen].

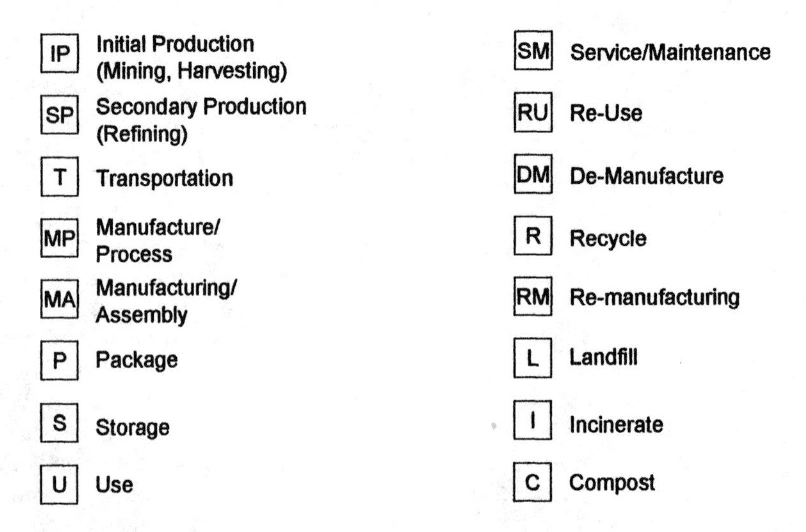

Figure 8.3 Sixteen stages for mapping product life cycle [Reprinted with permission from M. Sondeen].

product life cycles can become difficult to develop. However, it is a very worthwhile effort for the team to assemble as much of the information as they can to ensure a better understanding of the product's environmental impact. The bill of materials and the output from vendor selection should provide information on suppliers providing raw materials and product components. Marketing should have customer data stating where the products go, and the PT should know who hauls away waste and what happens to scrap on the manufacturing line. From this information, the teams can begin to map the product's life cycle.

Figure 8.4 shows a first pass at a team's mapping of a product life cycle. This figure shows that there are four outside parties associated with the company, each with a different role to play in the product's life cycle. There are three suppliers, A, B, and C that provide materials and components for the product. The materials from suppliers A and B are combined and processed to form a part that is then assembled with parts from supplier C to form the product. The manufacturing process produces two waste flows, one of which is recycled by reusing it in the manufacturing process, and the other flow is to the landfill, where it is classified as hazardous waste, costing the company a significant amount of money to dispose of properly. Once the product is made, it is packaged and transported to the customer.

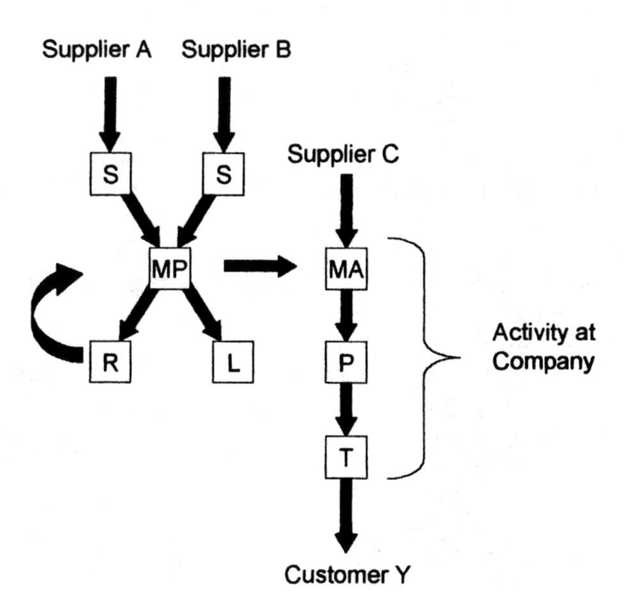

Figure 8.4 The team's first attempt at a life cycle map [Reprinted with permission from M. Sondeen].

The team has made a good start at mapping the life cycle of the product, but they must try and extend the map by defining the stages that are part of the supplier and customer chains. Suppliers A and B are both large international corporations and claim that due to trade secrets they cannot provide information regarding the types of materials that are in their products. As a result, the team is limited to the information they obtain from the material safety data sheets (MSDS). From this information, they must infer the origins of the materials that they can identify. Supplier C has recently implemented their environmental and money saving initiatives. As part of this initiative, they have recently initiated a reusable packaging initiative for their component shipments, in which they deliver components in their packaging and retrieve empty containers from the last shipment of parts. This new policy has benefits for both companies. Supplier C saves money on packaging, and our company does not have the disposal costs from the packaging material. Although our company has an open relationship with supplier C, they are unable to provide useful information about their suppliers' materials and sources. Therefore, the team will be unable to trace the materials from Supplier C back to their origins. The new information that the team has gathered from their suppliers can now be added to the map.

Our company provides components to customer Y that are used in their products, which are then shipped to the end user. Our company is an easily replaceable supplier, and when Y communicates with us, they generally provide only the engineering specifications for the product they desire. Our company is not involved in Y's product design or its design requirements. Once the product is delivered to Y, our company has no influence over its use or end-of-life fate. The team can also add what they know from customer Y to the map, which is redrawn and shown in Figure 8.5.

Now that the life cycle map has been defined in as much detail as possible, two issues arise. One, the map is incomplete, which is to be expected due to the complexity of the global industrial system. Two, even if a life cycle stage is known, obtaining the inventory information for this stage can be difficult if not impossible. Obviously, the most accurate and complete part of the map will be the company's portion. Following along this line of reasoning, it is then generally clear where the company can obtain the information it needs to make changes to the product to improve its environmental performance. This area where the company can make improvements is shown in Figure 8.5 by the dashed line, and is defined as the sphere of influence. This sphere of influence (SI) is where the team should focus its efforts on developing a life cycle focus and improving the product. As the company improves its own products and operations, its goal should be to expand the SI to include suppliers and customers where possible, so that eventually, the company can have an influence over their products' entire life cycles.

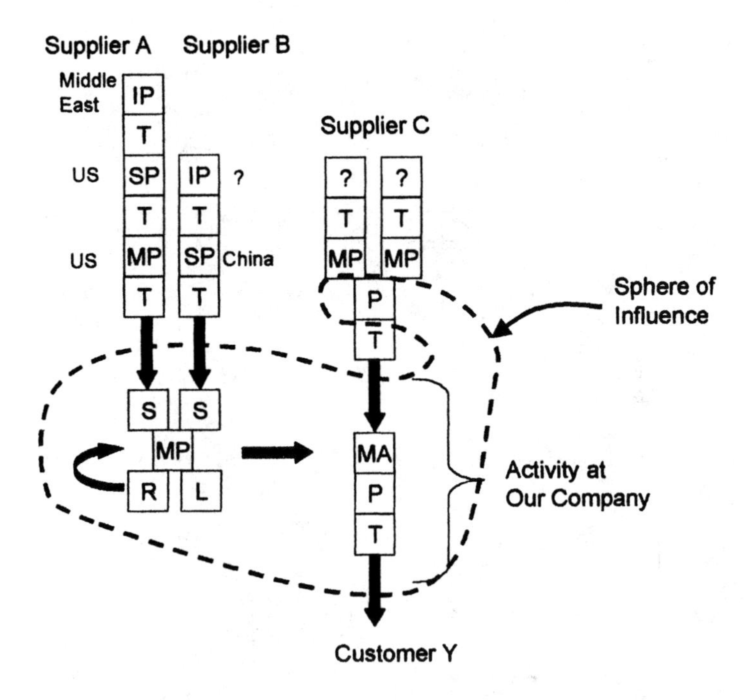

Figure 8.5 Life cycle map with supplier information depicted and sphere of influence defined [Reprinted with permission from M. Sondeen].

Inventory Tools

Inventory tools are used to determine the materials and energy that flow in and out of the various manufacturing processes. These tools include the process flow chart and an energy balance. The process flow chart is a diagram that shows an accounting of all the materials entering and leaving the processes on the manufacturing line. The layout of the line is generally used as the backbone of this tool, then the CPDT must analyze the materials going in and out of each process, the flow rates of these materials, and any losses that occur. This tool has been used for several years as part of pollution prevention assessments. Figure 8.6 shows a process flow diagram for the production of a metal safe. In addition to the materials, an energy balance should be performed, in which the energy going into a process and waste energy going out is measured to the best of the team's ability. All the gathered data should be organized and documented. It may be advisable to measure the flows by shift, day, or week, but it is useful to convert this number to a per product basis for comparison reasons. This information will be used in conjunction with assessment tools in the planning phase.

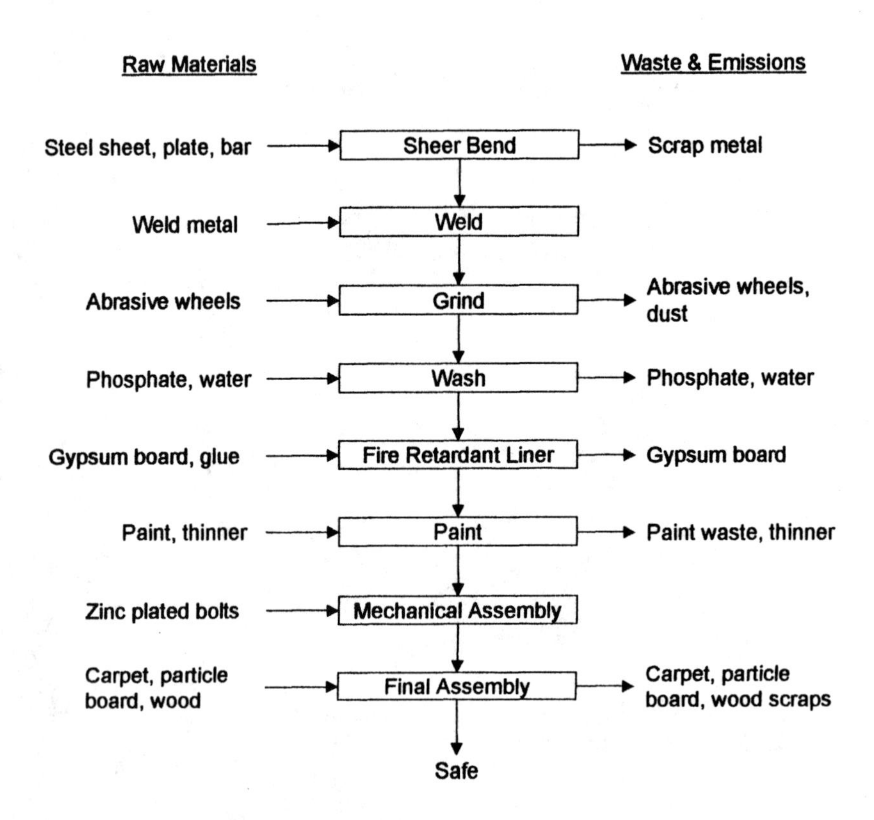

Figure 8.6 Process flow diagram for the manufacture of a safe [Reprinted with permission from M. Sondeen].

Voice of the Customer

The voice of the customer (VOC) is a useful way to seek out those product characteristics that interest the customer, and look for product improvements that can be incorporated into the next product design. There are two ways of gathering this information: 1) talk to customers and gather needs and desires for product and 2) examine field reports to look for product complaints and problems. In this phase of gathering data, we will concentrate on the field reports. Also be aware that the customer is not just the party buying your product, but also can be considered to be members of the surrounding community that have a stake in the burden of pollutants and the materials used in the factory. These concerns can present themselves through request for information, or through complaints that are made to local or state governments. The design team should be aware of regulations governing the use of materials and their release into the environment, and should be look ahead to future regulations that could be imposed. All of the information gathered in this step should be made available in the product planning phase of the new product.

End-of-Life

The end-of-life (EOL) of the current product should be considered and discussed. The team should understand where the current product ends up, before it can consider strategies for the new product. The information gathered in this step can be used in defining product requirements for the new product in the product planning phase. For example, if the company is moving toward re-manufacturing of some of its products, then this requirement needs to be examined early in the design process, and incorporated into the product specifications.

The most prevalent EOL for a product in the US is disposal in a landfill or incinerator. Following this option are: recycling, reuse, remanufacture, or a combination of these. Two main product characteristics are used to predict EOL fates: wear-out life and the technology cycle, These are shown in Table 8.1 [Rose, et al., 1998]. Additional product characteristics can also affect the EOL strategy of the product. These characteristics should be kept in mind by the CDT and CPDT when developing the new product. These additional characteristics include: design cycle, replacement life, reason for obsolescence, functional complexity, hazards, cleanliness of product, number and type of materials, number of modules, number of parts, and size [Rose, et al., 1998]. Probably the most useful tool that can help with design for EOL strategy is designing a more modular product, which was discussed at length in Chapter 4.

Table 8.1 Proposed end-of-life paths based on product characteristics [Adapted from Rose, C. M., K.A. Beiter, K. Ishii, K. Masui, "Characterization of Product End-of-Life Strategies to Enhance Recyclability," *Proceedings of the ASME Design Engineering Technical Conferences*, Atlanta, paper number DETC98/DTM-5742 on CD-ROM, 1998. Reproduced with permission from ASME.]

Wear-out Life	Technology Cycle	End-of-Life Path
Short	Long	Recycle materials
Short	Short	Re-use components or parts
Long	Short	Re-manufacture components or parts
Long	Long	Recycle materials

8.1.2 Project Planning Phase

The project planning phase in both the product and the manufacturing process development methodologies is the same. Although the planning phases for these methodologies are undertaken separately, the teams should be constant communication. This communication is critical for identifying the environmental needs as well as the product and process needs. Figure 8.7 shows the DFE enhanced planning phase, which includes the following tools: the streamlined life cycle assessment matrix, waste ratios, EOL strategy, VOC, and Quality Function Deployment (QFD). Each of these tools will be briefly described. At the end of this chapter is a reference list and a suggested reading list in which these tools are described in more detail.

Streamlined Life Cycle Assessment Matrix

A streamlined life cycle assessment (SLCA) is a qualitative analysis of the inventory data collected in the production and service phase. An SLCA is a much simplified technique following the same process as a life cycle assessment (LCA). An LCA is an attempt to evaluate the environmental impacts of a product or process in which the entire life cycle is taken into account. There is an attempt to quantify all energy and raw materials used and waste released, during the product's or pro-

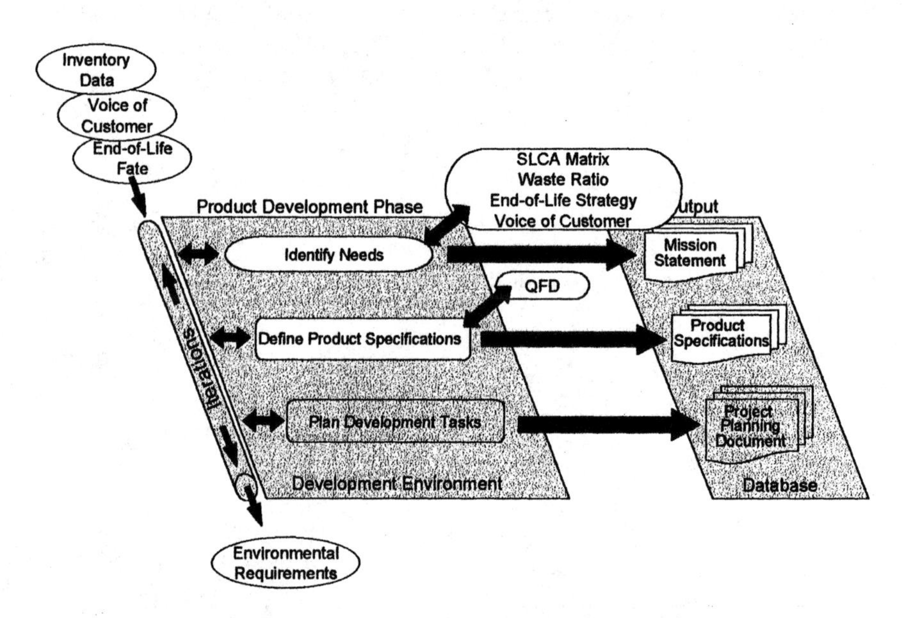

Figure 8.7 DFE enhanced project planning phase [Reprinted with permission from M. Sondeen].

cess' life cycle which includes raw material extraction, processing, manufacturing, transportation, distribution, use, and end-of-life. LCA's are quite expensive to undertake and can take years to complete. In addition, much of the data is incomplete, as discussed earlier. Furthermore, the results can be quite controversial, since experts in the fields of environmental science and toxicology have a limited understanding of the waste emission effects on the environment and in animals and humans. In a product design cycle, using an LCA approach is not feasible because of their expense and timeframe; therefore, a streamlined approach was developed.

Several methods have been used to streamline the LCA to make it more useful and practical for a design organization. First, boundaries are defined to limit the need for following all materials throughout the entire life, for which much of the data is not available anyway. Second, limits are placed on examining the impact of the use of materials and release of waste. In some organizations, only the amount of releases are considered, or the type of material released is examined if it is under regulation by the government. Another alternative is to consider only those impacts that are critical in the locality of the plant, such as water usage. Finally, in some cases, qualitative data is used when quantitative data is not available. Streamlining the LCA process is not an ideal method to assess the environmental impact of a product or process. However, the development teams can still gain valuable insight into improving the environmental performance of their products and processes through the use of this technique.

SLCA techniques have been developed in several companies including AT&T, Dow Chemical, and Rohm and Haas [Graedel and Allenby, 1995], however, the one we suggest using has been well-documented and described by Graedel [1998]. Table 8.2 shows a product assessment matrix in which each life cycle stage is evaluated based on a number of environmental stressor categories. Each category receives a score of 0, 1, 2, 3, or 4, in which 0 represents significant impact and 4 represents no known concern. This scoring may seem backwards, however, the authors felt that in a business setting an improving score should correspond to an improving product having less environmental impact. Each score should be accompanied by comments giving the reasoning behind the choice. When all ratings have been given, the product or process is given an overall score called the environmentally responsible product rating (R_{ERP}) [*Streamlined Life Cycle Assessment* by Graedel, © Adapted by permission of Pearson Education, Inc., Upper Saddle River, NJ]:

$$R_{ERP} = \sum_i \sum_j M_{i,j} \qquad \text{(Equation 8.1)}$$

Where $M_{I,j}$ represents a matrix element in row i, column j. The scores of all the matrix elements are summed to give the product rating, with a maximum score of 100 possible. This rating can be used to compare various design options.

An SLCA matrix should be used in the identify needs design step to perform a qualitative analysis of the inventory data. The most logical place to draw the boundaries for this analysis is using the sphere of influence. The life stages that are inside the SI are put in the left column. The CDT and CPDT should work together to complete the SLCA's for the product. The first few times an SLCA is used, the company may have to bring in additional members to the team, such as a health and safety personnel. The scores should be carefully documented and justified. This matrix is explained more fully in Appendix C. After its completion, the SLCA can be used to develop product requirements that seek to reduce the impacts caused by the product's design, manufacture, use, and disposal. Although this process is highly qualitative and subject to the team's judgement, the process can yield valuable information that can be used to guide the development effort in improving the environmental improvement of the company and its products.

Table 8.2 SLCA matrix [*Streamlined Life Cycle Assessment* by Graedel, © Adapted by permission of Pearson Education, Inc., Upper Saddle River, NJ]

	Environmental Stressor				
Life Stage	Materials Choice	Energy Use	Solid Residues	Liquid Residues	Gaseous Residues
Premanufacture	1,1	1,2	1,3	1,4	1,5
Product Manufacture	2,1	2,2	2,3	2,4	2,5
Product Delivery	3,1	3,2	3,3	3,4	3,5
Product Use	4,1	4,2	4,3	4,4	4,5
End-of-Life	5,1	5,2	5,3	5,4	5,5

Waste Ratio

A waste ratio is a measure of efficiency developed by 3M that can be used to indicate a base level of performance, and then be used to measure progress. In the last phase, energy and material usage measurements should have been taken on the manufacturing lines. From these numbers the waste ration can be calculated:

$$Waste\ Ratio = \frac{Waste}{Product + Byproduct + Waste} \qquad \text{(Equation 8.2)}$$

All these numbers should be in the same units, such as pounds. This number can highlight the most inefficient processes on the line and those needing the most improvement. The team should estimate raw material cost differences between the current process, and the best practice based on feasible technology in the marketplace.

End-of-Life Strategy

The EOL fate of the current product was assessed in the production and service phase. At this point in the planning process, the teams should identify the strategy that they wish to follow based on the product characteristics given in Section 8.1.1.4 and discussed below.

- Technological Cycle
  What is the technology cycle of this product? How long can it be considered desirable by the marketplace? This evaluation should be done in conjunction with marketing, which will have an understanding of the marketplace and the competitors.
- Product Life
  What is the estimated life of the product? What is the limiting factor on the product's life? Products may be discarded for many reasons which can include: wear, obsolescence, desire for technology upgrade, and loss of utility or function.
- Design Cycle
  What is the design cycle of the product? How often are new models designed before a technology upgrade is made to significantly change the product design from being in an evolutionary design cycle to being in an original design cycle?

Once these questions have been answered and their relationships assessed, the team must decide upon an EOL strategy for the new product. Should the EOL strategy of this new product differ from the old, and is a change possible? In the case of a company providing a part or component that resides in another company's product, changing the EOL strategy may be infeasible. If the product goes directly to the end user, the company may be able to develop a take-back strategy that will allow them to receive old products and reclaim the parts and materials for re-use or recycling. Figure 8.8 shows a chart than can help the CDT in evaluating product characteristics and formulating a new EOL strategy. For example, the team may be able to develop field replaceable units that can extend the life of the product which may provide a technology upgrade or may replace a component that has a significantly shorter life than the rest of the product.

Voice of the Customer

The voice of the customer (VOC) gathered during the last design phase is used to help identify the needs of the new product. Recall that in the last phase we relied on the field reports to do this. In the next design step, we will rely on customer contact to identify new product specifications and requirements.

Quality Function Deployment

Quality function deployment (QFD) is the process of following the steps to gather and translate customer needs and desires into the house of quality (HOQ). The

HOQ was discussed in detail in Chapter 4, so we will only briefly discuss it here. In gathering customer information, the team should include questions about environmental desires, requirements, and trade-offs. Furthermore, the HOQ can be valuable through benchmarking the current product against competitors' products, which may be easier to disassemble and separate for recycling, or it may contain fewer materials. In any case, the HOQ is now undertaken with the additional focus of including environmental requirements.

After the HOQ is complete, the team must develop the product and process specifications. Some of the product specifications that may arise due to environmental considerations could include the reduction of the number of fasteners, the reduction of material variety, or the substitution of materials. The CPDT should consider the manufacturing process, testing, and packaging specifications addressing environmental requirements such as increasing energy and material usage efficiency, reducing waste, and substituting for hazardous or controlled materials. At the end of the planning phase, the teams prepare the milestone report as discussed in earlier chapters. The justification of the environmental specifications and requirements should be included with the rest of the product and process specifications. The output from these design steps are then used in the next phase, conceptual design.

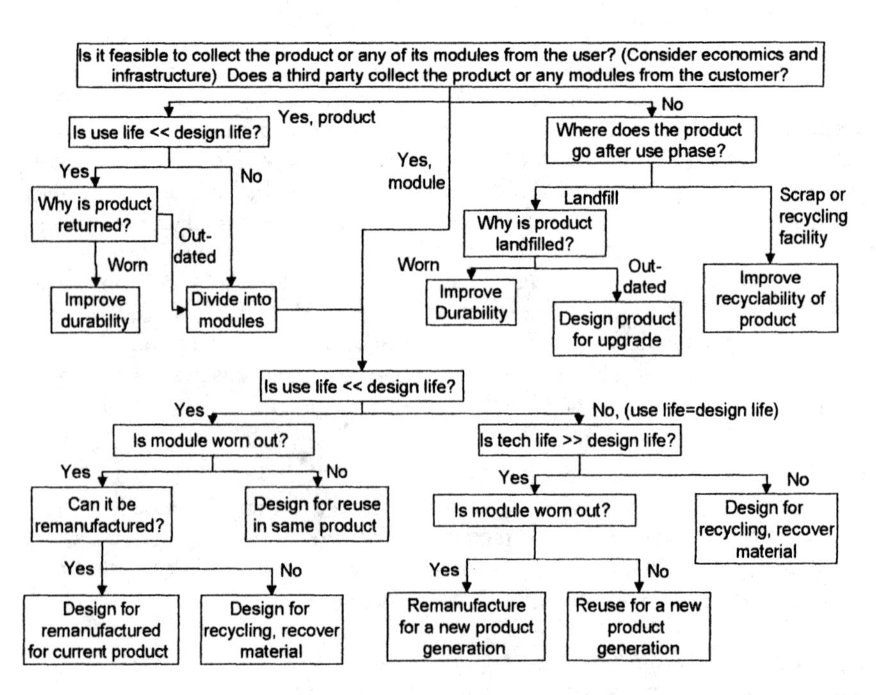

Figure 8.8 EOL planning tree phase [Reprinted with permission from M. Sondeen].

8.1.3 Conceptual Design Phase

In the conceptual design phase, improvement tools will be used to help meet the specifications developed in the planning phase. Because the product and process phases are somewhat different, they will be discussed separately.

Product Conceptual Design Phase

There are three DFE tools that are incorporated into the conceptual design phase for the product development model: design for modularity, concept selection matrix, and the concept comparison matrix. Figure 8.9 shows the incorporation of these DFE tools in the conceptual design phase. Although the DFE improvement tools are used in only two of the design steps, as shown by the highlighted outputs, they have an effect on three of the five design step outputs. Next, these three DFE tools will be discussed individually.

Design for Modularity In the design step, generate concepts, the CDT uses the environmental specifications along with the product specifications to begin the task of developing product concepts. The team needs to generate several product concepts, then rationally compare them to determine which best meets the specifi-

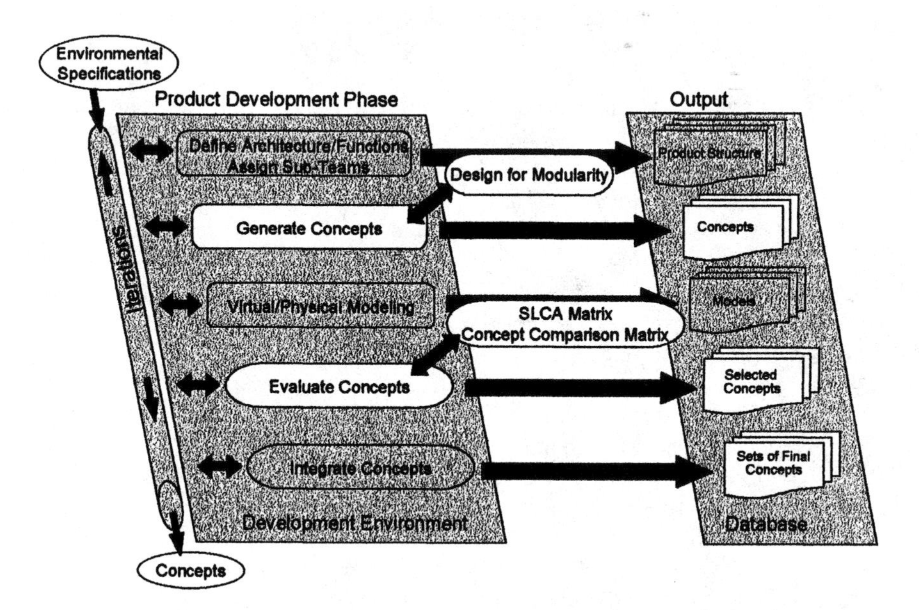

Figure 8.9 DFE enhanced conceptual design phase for the product development model [Reprinted with permission from M. Sondeen].

cations, which should include what the intended end-of-life fate is for this product. Note that a product with a modular architecture is typically easier to recycle, reuse, or remanufacture, which are characteristics that need to be considered when defining the product's EOL strategy. In Chapter 4, we discussed how to assess a product's modularity which is undertaken in the define architecture/functions design step. Using this information, the CDT needs to align the new product's architecture with the EOL strategy decided upon during the project planning phase. It may turn out that the product is complex enough that various modules and components could have different EOL strategies. In this scenario, the product is usually disassembled and the different modules and components are handled according to their EOL scheme. The EOL schemes that the design team should consider are:

- Recycle
  A product or module that is to be recycled, ideally, should be made of a single material. These modules should be easily separated from the rest of the product. Other means of recycling components that consist of multiple components do exist, and the team must explore these options if it is infeasible to make a single material module, which is the case more complex modules such as computer wiring boards. In these more complex components, the part is shredded, and the materials are sorted according to physical properties such as buoyancy, mass, or magnetic attraction.
- Reuse
  A product or module that is to be reused must be durable and easily removed from the product without damage. Its removal must be economical; therefore, the CDT must consider its location within the product and its connections to other modules. These modules can be valuable to the company for re-use in other products or as repair units.
- Remanufacture
  A product or module that is considered a candidate for remanufacturing is usually those with high cost and a long technology cycle compared to the wear out life. Xerox Corporation has made a commitment to the remanufacture of its copiers, thereby saving millions of dollars in raw material costs. Products or modules that are to be re-manufactured must be examined for ease of removal. Therefore, these components must be easily accessible and easily disconnected from adjoining modules. Products that are to be re-manufactured must be designed for ease of disassembly. These design techniques are beyond the scope of this book, but the suggested reading list includes materials that discuss design for disassembly in detail.

Once the CDT has generated a number of product concepts, then the task of evaluating them must be undertaken. A quick evaluation is done in the concept generation design step, as discussed in Section 4.2.2, using the concept selection matrix shown in Figure 4.11.

Streamlined Life Cycle Assessment Matrix A useful tool in evaluating the various concepts is the SLCA matrix. Typically, at this stage of the design process, the best two or three concepts have been selected and a more in-depth evaluation is performed. If the team has time, they can develop SLCA matrices for these two or three concepts, then these can be used to assist in deciding upon a final concept. The SLCA matrices for the new concepts should be compared with the current product's matrix to ensure that the new concepts represent an improvement.

Concept Comparison Matrix In most cases, a company must balance its commitment of environmental improvement with cost and quality requirements. The product must also meet the market needs. If a design meets the environmental goals of the company, but is not cost competitive or does not meet customer quality requirements, then it likely will not sell.. Customers state that they want greener products and will pay extra for them, but numerous studies have shown that in reality cost and quality desires override environmental concerns. Therefore, it is important to consider all three of these characteristics when evaluating the best concepts.

A useful tool for evaluating concepts for cost, quality and environmental considerations is the concept comparison matrix, shown in Figure 8.10. The current product concept and the new concepts are entered in the rows on the left. The columns contain the evaluation criteria. The most important quality requirements are input into the matrix from the product specifications or from the house of quality completed in the planning phase. Cost criteria that can include material cost, manufacturing cost, and sales price can be used in the cost section. The environmental criteria are those that were defined in project planning and deemed the most important. Each concept is then evaluated on a scale of 1 to 10 and the score is placed in the appropriate row and column, as shown by the letters A, B, and C. The CDT and CPDT teams with input from the management team can assign weights to the quality, cost and environmental categories, as shown by X, Y, and Z. Importance weights are also given within the three categories of quality, cost, and environmental considerations, as shown by the letters L, M, and N. Finally, each concepts is given a satisfaction score which is calculated using the following formula [Adapted from Zhang, Y. H., P. Wang, C. Zhang, "Life Cycle Design with Green QFD-II," *Proceedings of the ASME Design Engineering Technical Conferences*, Atlanta, paper number DETC98/DTM-5719 on CD-ROM, 1998. Reproduced with permission from ASME]:

$$S_3 = X(A_1L_1 + A_2L_2 + A_3L_3) + Y(B_1M_1 + B_2M_2 + B_3M_3) + Z(C_1N_1 + C_2N_2 + C_3N_3) \qquad \text{(Equation 8.3)}$$

This calculation is shown for concept 3. This satisfaction score allows the team to

	Quality			Environment			Cost			
Requirements / Product Concepts	Criteria Q1	Criteria Q2	Criteria Q3	Criteria E1	Criteria E2	Criteria E3	Criteria C1	Criteria C2	Criteria C3	Satisfaction
Benchmark										
Concept 1										
Concept 2										
Concept 3	A_1	A_2	A_3	B_1	B_2	B_3	C_1	C_2	C_3	S_3
Weights	X%			Y%			Z%			
Importance	L_1	L_2	L_3	M_1	M_2	M_3	N_1	N_2	N_3	

Figure 8.10 Concept comparison matrix [Adapted from Zhang, Y. H., P. Wang, C. Zhang, "Life Cycle Design with Green QFD-II," *Proceedings of the ASME Design Engineering Technical Conferences*, Atlanta, paper number DETC98/DTM-5719 on CD-ROM, 1998. Reproduced with permission from ASME].

look at how each of the concepts compares in satisfying the three requirements of quality, cost and environment. This matrix allows the teams to consider all concepts and evaluate them consistently. As we have seen earlier in Chapters 4 and 5, the team may find that it can produce a better product by combining several features of two different concepts, thereby generating another option. Then, the team must iterate through the design steps to develop this new concept's model and evaluate it against the others.

Process Conceptual Design Phase

The manufacturing process development model, as we saw in Chapter 5, is similar to the product development model. In the conceptual design phase, there are five design steps and outputs, with four of the five outputs affected by the DFE tools and input from previous phases. Three DFE tools are used in this phase: the process interaction matrix, the SLCA matrix, and the concept comparison matrix. The other two process models, test and packaging development models, are similar and

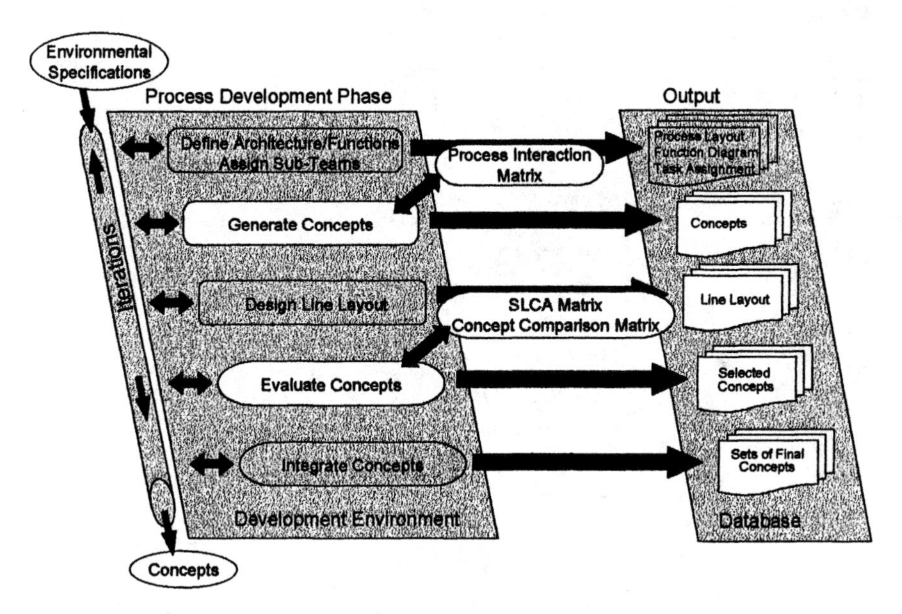

Figure 8.11 DFE enhanced Conceptual Design Phase for the manufacturing process development model [Reprinted with permission from M. Sondeen].

use the same DFE tools. Therefore, we will only show the manufacturing process model and discuss the tools used in conjunction with this model.

From the planning phase, the environmental specifications are included with the other process specifications. The CPDT should incorporate these specifications into concept generation and design of the line layout. The team should have set environmental goals when defining the process specifications;, now they need to generate concepts to meet these goals. They may be able to meet these goals by implementing in-house recycling, material substitution, or eliminating process steps. However, the team must also be open to replacing existing technology with new "clean" process technology. For example, if the company manufactures printed wiring boards, it should consider replacing CFC-based solvents with water-based solvents. In general, only one or a few of the processes are redesigned to meet the project goals, assuming that much of a line is re-used. In developing these new concepts for the manufacturing line, testing, or packaging, as the case may be, the changes in one step could affect other steps. For example, switching to a water-based coating process may require that a more caustic cleaner be used in the preparation station. The process team needs to be aware of possible interactions between processes, therefore a process interaction matrix is used, which is shown in Figure 8.12. In this matrix, the team needs to evaluate the interaction

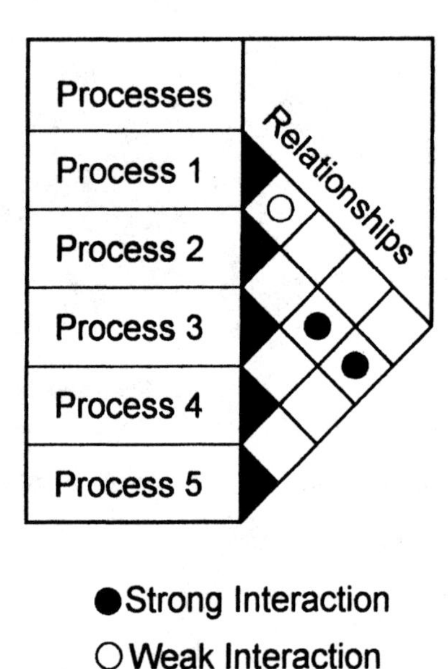

Figure 8.12 Process interaction matrix [Reprinted with permission from M. Sondeen]..

effects between processes. All interactions need to be examined, but especially those having strong interactions. These interactions must then be taken into account when evaluating and selecting components, and designing the line layout.

The other two DFE tools, the SLCA matrix and the concept comparison matrix, are used in the evaluate concepts design step. Both these tools were discussed in the last section, therefore we will only discuss how they can be used in a process setting. The SLCA matrix should be used to evaluate the best two or three concepts with the current process to ensure that there is an environmental improvement, or at the least, there is no degradation of performance. The concept comparison matrix shown in Figure 8.10 is used to evaluate process, test, or packaging concepts. The criteria may differ; for example cost criteria would include capital expenditure, payback period, and maintenance costs. Otherwise, the process is identical to that followed in the product design model.

8.1.4 Design Phase

At this point in the design process most of the DFE effort is complete. In the design phase, only general considerations need to be made and no new tools will be used. When finalizing the design of the product, the CDT should consider the following issues:

- Product and component mass should be minimized without compromising strength or durability.
- Avoid selecting materials that require extensive machining operations to achieve a desirable product finish, which would require high energy use.
- Design parts that can be manufactured with low energy process methods. For example, machining typically uses more energy than forging.
- Minimize part count. Many functions can be often be incorporated into a single part, however, this requirement has to be balanced against the desire for modularity, which may call for more parts.
- Eliminate material finishes or additives that can contaminate the material so that it cannot be recycled.
- Mark all plastic parts with ISO identification symbols for recycling purposes.

The issues addressed in process design are a bit different than those considered in product design. The CPDT should consider the following issues in the design of the manufacturing process, test or packaging lines:

- Select the most energy efficient equipment,
- Size equipment conservatively, yet select equipment sufficiently oversized to meet the anticipated needs of future production numbers, and
- Select equipment that has long life, good reliability, and easily maintained.

In the embodiment design step in the product design model and in the vendor selection and approval design step in the process models, the CDT and CPDT select materials, components, and equipment from suppliers with help from the procurement organization. In this selection several considerations that should be taken into account.

- Distance from company

  Transportation of parts can have a large impact on the environmental performance of a product, thus the reduction of shipping distances can reduce this impact. Therefore, local vendors should be given a priority. There are side benefits to having local suppliers including close vendor relationships, tighter control over quality, just-in-time delivery of parts, and quick response to quality issues.
- Environmental management program

  The supplier should be questioned as to the existence of an environmental policy. The company should ascertain if they have pollution prevention or waste minimization policies in place. These policies show a proactive stance regarding the environment, and a willingness to work with your company to implement these types of programs for the parts they supply to you. A supplier with an environmental management program is also more like to have a quality program in place such as ISO 9000. Finally, a company that has quality and environmental policies in place is likely to be better organized and managed than a company that does not have these programs.

In addition to these general considerations, there are considerations for product design in particular. These product design considerations include:

- Material content
  It is desirable to work with a supplier that is willing to share the material content of their products. The more information that the CDT has about component materials, the better informed their decisions can be in the face of improving the environmental performance of their own products.
- Packaging
  The supplier should be engaged in the activity of reducing their packaging without reducing the protection of their components. The team may be able to enter into a relationship with the vendor to reuse packaging materials, thereby reducing waste in both companies. This type of relationship is most easily accomplished with local suppliers.
- Component characteristics
  The supplier's components should be designed to minimize energy use, and facility reuse, remanufacture, or recycling. The vendor should also incorporate recycled materials or remanufactured materials in their parts.

The CPDT team should consider the following issues:

- Equipment efficiency
  The equipment being considered should be energy efficient, use state of the art clean technology, and avoid the use of controlled or hazardous materials.
- Maintenance and remanufacture
  If desired, the team should consider those companies that offer maintenance service and remanufactured, replacement parts backed up by warranty. Finally, the team needs to consider what will happen to the equipment once it is not longer needed, will the vendor reclaim the equipment or is there a used market for this type of equipment.
- Equipment durability
  The team needs to consider the equipment's expected life in relation to the technology life cycle. Also, they need to consider the equipment's expected mean time between failure record and the type of service and time required to get the equipment back on-line.

- Upgradability
  Equipment that can be upgraded to take advantage of new technological advances, expanded capacity or increased functionality is more desirable than equipment that cannot be upgraded. This type of equipment may cost more, but the benefits may very well outweigh the cost differentiation.

It is unlikely that a single vendor can meet all these criteria, but one that can meet a few of these criteria should be favored over those that cannot. A close relationship with the suppliers can help guide the CPDT in the design of the process equipment and lines, thereby helping improve the environmental performance of the factory.

8.1.5 Production Preparation Phase

The production preparation phases for the product and process development models are quite different. However, there are only two DFE activities in this phase, and they are associated with the procurement and the pilot production steps, which are present in all the models. Associated with the procurement step is the DFE task of expanding the sphere of influence (SI). The SI map was first developed in the production and service phase of the current product. Now, the new product and process designs are complete and near to production. The SI map should be revisited to see if the team can expand the boundaries of the map.

In the last section, we discussed the inclusion of suppliers in our environmental requirements through inclusion in the design team and developing reusable packaging. These type of relationships expand the SI, and thereby improve the environmental performance of the product. The CDT and CPDT should seek out those suppliers who are willing to share information and are committed to environmental improvement. In addition to the purchase of materials, components and equipment, the team should also consider the waste streams. Wherever possible, the teams should seek to expand the SI by identifying secondary markets for their waste materials. For example, a small company, E Media, was able to procure recycled material for their video cassette tapes from companies that made disposable diapers [Carlson-Skalak, et al., 2000]. This relationship benefited both companies in that the waste from the diaper making process did not go to the landfill, but resulted in a financial gain for that company rather than a disposal cost, and E Media benefited from lower raw material costs. The design teams should actively seek these types of mutually beneficial relationships.

In the pilot production step, the CDT and CDPT should ensure that the product manufactured, tested and packaged on the manufacturing line meets the product and process specifications defined, which include the environmental requirements. At this time, the teams should assess whether they have accomplished the environmental goals that they set out for themselves by comparing the new product with the product they evaluated during the production phase at the start of this design cycle.

8.1.6 Production and Service Phase

In this phase, the product team (PT) needs to monitor the product and process and ensure that it continues to meet the goals defined by the teams earlier in the design process. All goals need to be assessed to determine if the project is a success, and what lessons were learned and can be passed onto the next project's development teams. Documentation is key, as has been discussed throughout this book. Documentation of the work and results will facilitate the inventory process for future projects and development teams.

8.2 DFE EXAMPLE

Now, an example using the DFE methodology just described will be discussed. This methodology was implemented within a small company that was interested in redesigning a portion of their manufacturing process. First, we will describe the company and their products, an overview of the project that they were undertaking, and the implementation of the DFE methodology. This methodology will only be discussed through the conceptual design phase, since at the time of this writing, that is where the project stands; therefore, we are unable to provide final project results.

8.2.1 Company Background

The small company with whom we worked builds safes. At the time we worked with them, they were looking to accomplish several goals. First, they wanted to gain a stronger presence in the marketplace and expand their business. As part of this goal, they needed to improve manufacturing operations, decrease costs, and improve delivery times. The owners were also interested in improving the environmental performance of their company, and reducing the impact of their product. The safes the company manufactures are available in two models. The basic model is finished in a rough-textured paint. The higher end model uses thicker steel plate and is finished in a smooth, high-gloss paint. Several options are available on both models including a gypsum liner for fire protection, various interior shelving configurations, and various paint colors.

The safe production process is shown in Figure 8.13. The process begins with incoming sheets of raw steel that are sheered to size. Next, holes are formed with a metal punch, and the pieces are bent in a press brake to form the body of the safe. The parts are then welded together to form the body and the door. The welds are then ground to remove surface irregularities and gives a smooth finish. The safes are washed with a phosphate-based detergent to remove dirt, oil, and grinding particles, and to etch the surface for better paint adhesion. Next, the safe receives a gypsum liner if ordered. The safes are then primed and sanded to ensure a high-quality paint surface. The safes are painted in a booth with a hand-operated spray system, and then baked to cure the paint. The door is then attached to the safe body, and the locking mechanism and interior trim are installed. Finally, the safes are wrapped in foam and corrugated cardboard, and placed on a wood pallet for shipping. The product is distributed through a dealer network, who sells to end users.

8.2.2 Project Overview

The focus of this DFE project was to redesign portions of the production process to improve production flexibility, reduce costs, and reduce environmental impact. A team was formed to accomplish these goals and included the author, two graduate research assistants from the University of Virginia, the owners of the company, the plant foreman, a manufacturing engineer, and an environmental engineer. The team then followed the procedure outlined at the beginning of this chapter. The following sections will overview the activities and describe the results obtained thus far.

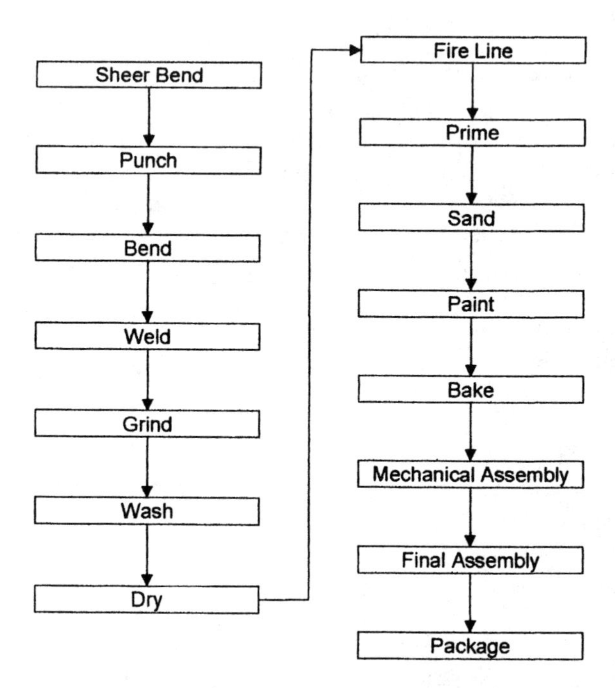

Figure 8.13 Process flow diagram of a safe [Reprinted with permission from M. Sondeen].

8.2.3 Production and Service Phase

The project team began by performing an inventory of the current line of safes and their manufacturing processes. The team started by examining the bills of materials for each safe model to compile a list of input materials. The most popular safe model contains over 140 different parts. Purchase order databases provided a list of suppliers, which numbered more than 40. Accounting for the waste in the process proved to be more difficult. The manufacturing engineer and plant foreman were interviewed, along with factory floor personnel to determine what wastes were produced during a safe's manufacture. Waste removal costs were lumped in with overhead costs, and not accounted for on a per product basis. Therefore, waste quantities had to be estimated from production volumes and monthly waste volumes. The energy estimates also proved to be difficult to obtain. The company's electricity records did not break down the usage between the manufacturing equipment and general office and utilities. Furthermore, the team had no measure of energy usage on the individual process machines. These numbers were not included in this analysis due to time constraints, and will be pursued at a later date. The transportation data was also difficult to obtain. Raw materials and components arrive via diesel truck. The impact of the deliveries was difficult to ascertain. The finished safes are transported via diesel trucks to distributors all over the United States, and finally to the end user. The safes can weigh over 1,000 pounds each, so a single safe requires a significant amount of energy to deliver. The only method to quantify the impact of the transportation process was to estimate the distance traveled for these shipments.

From this data and a process flow diagram shown in Figure 8.13, the team then developed a life cycle map. It soon became clear, that with over 140 parts and 40 vendors, the map would soon become unwieldy. As a result, the team evaluated the raw materials and component parts to ascertain those most likely to cause the greatest environmental concerns. The following criteria were used for this initial screening:

- Significant disposal costs,
- Non-regional supplier,
- Low product to waste ratio,
- Use and/or disposal regulated by government organizations such as the EPA, OSHA and NIOSH,
- Significant social impact on the local community,
- Significant health risk to employees,
- Significant environmental impact, and
- Low or zero recycled content.

The team evaluated each part, a portion of which is shown in Table 8.3. Those parts receiving a higher score were included in the life cycle map. The resulting map is shown in Figure 8.14 with its sphere of influence shown by the dashed

Table 8.3 Results of preliminary material screening [Reprinted with permission from M. Sondeen].

L3014 Safe		Disposal		Scrap	Non-regional	Zero Recycle	Env/Soc	
Part	Vendor	Costs	Regulated	Disposal	Supplier	Content	Impact	Total
Paint, Black	Paint Co.	X	X	X		X	X	5
Thinner, Urethane	Paint Co.	X	X	X		X	X	5
HR Bar Flat 0.37	Steel, Inc.							
Lock Combo S&G	Lock Co.				X	X	X	3
Snap Cap	Products, Inc.				X	X		2
Fire Shield 0.625	Gypsum Co.	X		X		X		3
Felt	Felt Co.				X			1
Phosphate Wash	Phosphate Supplier			X		X	X	3
Carpet, Taupe	Carpet Mills	X		X	X	X	X	5
Fiber Board	Fiberboard Co.	X		X			X	3

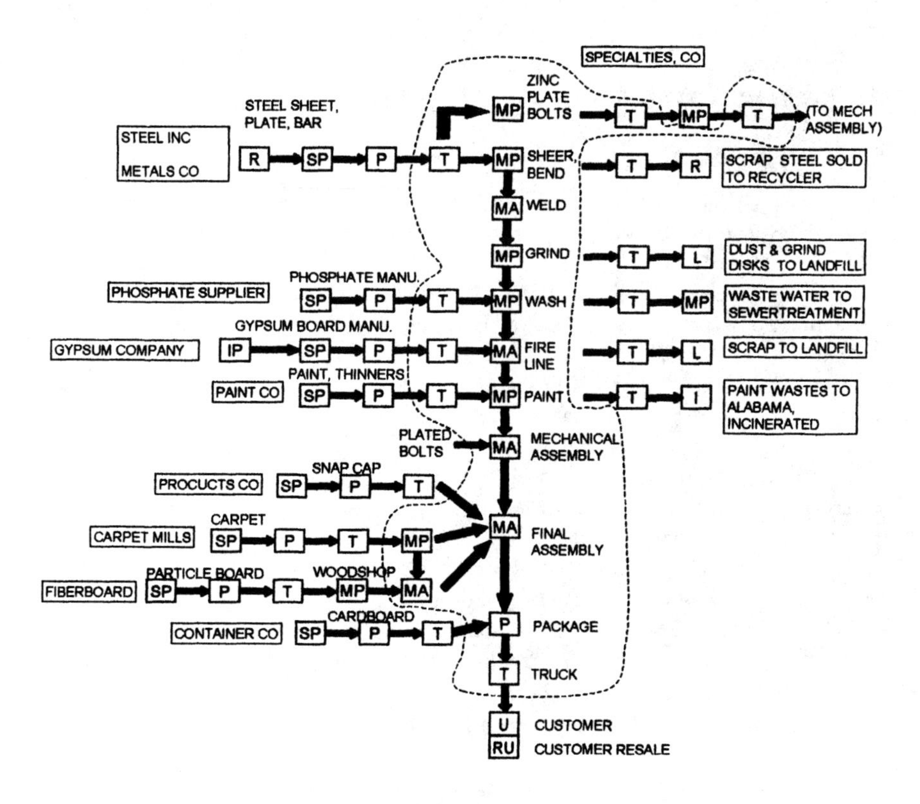

Figure 8.14 Life cycle map with SI defined [Reprinted with permission from M. Sondeen].

line. The SI is drawn with narrow boundaries for two reasons. One, this is the first DFE project undertaken by the company, and two, joint ventures with suppliers had never been attempted.

There is no particular end-of-life plan for the safe. However, sales and marketing was able to provide insight into the use and end-of-life fate of the product. The safe is generally kept by the original owner for several decades, as it is usually a one-time purchase. There are no parts that wear out over the life of this safe, and the product is usually replaced only in the event of a fire, burglary, or a desire to upgrade to a newer model. Damaged safes are either sold for scrap or sent to the landfill. If a safe is replaced through an upgrade, then the original safe is usually resold.

As for customer input, there have been no expressed desires for a more environmentally sensitive product. However, manufacturing personnel expressed a desire to know what hazardous materials are used on the line and their effects. Currently, the state regulators consider the company to be a small quantity generator of hazardous wastes, so regulation is minimal at present. However, management believes that stricter waste regulation is imminent.

8.2.4 Project Planning Phase

With the assessment of the current product lines completed, the team began work on identifying project needs. For this first project, management decided to concentrate only on the manufacturing process and to ignore the product design for this project. The team performed an SLCA, evaluating the life cycle stages within the defined SI. This resulting matrix is shown in Table 8.4. The matrix shows that the greatest impacts (the ones with the lowest scores) were: wash, paint, and transportation. The team elected to ignore transportation, and concentrate on wash and paint, which they can impact. The main concern regarding the wash station was the acidic, phosphorous-based detergent and its effect on regional waterways. The wash system drains to the local sewer, and on to the water treatment plant. The plant treats the water to lower the acid, but any particles containing lead from the safes and the phosphate detergent were not targeted in the treatment. The treated water was then released into a local river, which is part of a watershed. The painting process required the use of paints containing volatile organic compounds (VOCs). During the painting process, the majority of the fumes were vented to the atmosphere. Solvents containing methyl-ethyl-ketone (MEK) were used to thin the paint and clean the equipment. MEK is a concern for the health of the workers, and the paint waste is a controlled substance that must be handled as hazardous waste at a significant cost to the company.

Next, the current product's end-of-life fate was examined, and the following characteristics were determined:

- Wear-out life: very long, parts have an almost infinite life,
- Reason for disposal: fire damage, burglary, or decision to upgrade to different model,

- Technology cycle: relatively long, basic technology remains unchanged from year to year, and
- Design life: the product is typically redesigned every few years with those changes made to physical appearance and cost reduction.

Based on these characteristics, the team decided that the EOL fate, which was primarily resale or recycling, except in cases of severe damage where the safe is landfilled, was appropriate. There was no justifiable need to change the EOL of this product. In later redesign projects, the company may want to look at increasing the recycle content, but no other changes were needed.

At this point in the process, the team developed a set of project needs. From an evaluation of the inventory data, the SLCA matrix, and the EOL fate, the company decided to concentrate on the processes of wash and paint. When improving these processes, the company must take into consideration the voice of the customer, functional improvement, product quality, and cost reduction. The following is a list of project needs:

Table 8.4 SLCA matrix of current safe line [Reprinted with permission from M. Sondeen].

	Envrironmental Stressor						
Life Cycle Stage	Health Hazards	Energy Use	Solid Residues	Liquid Residues	Gaseous Residues	Environmental Externalities	Totals
Incoming transportation	3	2	3	2	1	1	12
Incoming packaging	4	2	1	4	4	3	18
Sheer	3	2	4	4	4	3	20
Punch	3	2	4	4	4	3	20
Bend	3	1	4	4	4	3	19
Weld	2	1	4	4	4	3	19
Grind	2	1	2	4	4	3	17
Wash	3	2	2	1	3	1	12
Fire line	3	3	2	4	4	3	19
Prime	2	3	2	4	4	2	17
Paint	0	3	2	1	1	1	8
Bake	3	2	4	4	1	2	16
Mechanical Assembly	3	3	3	4	3	3	19
Final Assembly	3	3	3	4	3	3	19
Package	4	4	3	4	4	3	22
Transportation	3	2	3	none	1	1	12
Use	4	4	4	4	4	4	24
Totals	48	43	49	58	51	41	290/432

Key:	
	4: No concern
	3: Minor concern
	2: Moderate concern
	1: Significant concern
	0: Extreme concern

- Reduce or eliminate health hazards associated with paint handling,
- Reduce or eliminate paint waste,
- Investigate alternative wash detergents and methods to avoid or eliminate discharge of detergent and waste into sewer,
- Cost justify all changes to process, and
- Product design is to remain fixed unless process changes warrant modification

Management then reviewed these needs and approved them.

8.2.5 Conceptual Design Phase

In the conceptual design phase, the team decided to leave the line architecture unchanged and addressed only the two stations that had the most environmental impact: wash and paint. In the wash station, the first concept considered was switching to a different detergent, such as a citrus-based degreaser. However, the phosphoric acid in the current detergent is needed to etch the surface of the safe to provide a surface resulting in a strong paint bond. Citrus-based cleaners would clean the pickling oils from the safe, but would not etch the surface. Next, the team considered a wash water reclamation and filtering system, which would reduce water usage and sewer discharge. Equipment is available for this type of application, and therefore this concept was considered as the most feasible. As for the paint station, the best concept is to use powder coating. In powder coating, the safe and the powder are given different electrical charges so that the paint powder is attracted to the safe and adheres to its surfaces. This paint has no VOC emissions and no solvents are needed for cleaning. Furthermore, there is significantly less waste, and what waste there is can be baked into an inert brick that can be safely landfilled. Hazardous waste handling would be eliminated. Finally, color changeovers could be made more quickly, as the equipment does not have to be cleaned between color runs as in the liquid paint system. Next, the team developed the process interaction matrix to look at the impact that changing these two processes could have on other stations. The team suspected that changing the paint technology could require changes to the surface preparation stations: wash, prime, and grind. In addition, recycling the wash water could impact the quality of surface preparation and affect the paint quality. These impacts are shown in the process interaction matrix in Figure 8.15.

Next, the team evaluated the new concepts using the concept comparison matrix, taking into account environmental goals, quality and customer requirements, and costs. The new concepts were evaluated against the current practices. The results of this comparison shown in Table 8.5, which is explained more thoroughly in Appendix D. Management decided that weights of 40% cost, 40% quality, and 20% environment were reasonable. The results from this analysis revealed that for the paint station, the paint quality demanded by the customer on

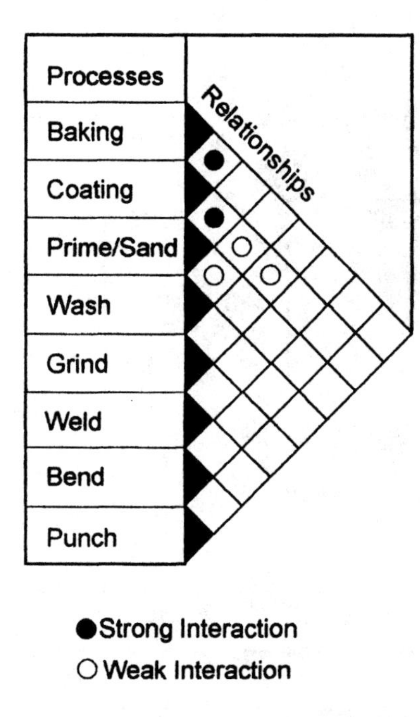

Figure 8.15 Process interaction matrix for process changes [Reprinted with permission from M. Sondeen].

the high-end safe was of a high-quality, high-depth finish, similar to that of an automobile. There was concern that the powder coating would not be able to duplicate the high shine finish. However, the switch to powder coating could significantly reduce waste disposal, raw material, and labor costs. Furthermore, several vendors were able to closely duplicate the finish desired on the high-end safe. This changeover is one that should be investigated further. For the wash station, although the reclamation system decreased environmental burdens, it also came at a high cost. The long pay back period could not outweigh the environmental improvements.

At this point in the process, the company is actively investigating the changeover to powder coating. They are working with several equipment and paint powder manufacturers to understand the technology, its costs, its advantages, and its trade-offs in comparison to the conventional paint system now in use. The are testing to determine what powder formulations will give them the high-end finish they desire. Furthermore, the integration of the new paint technology with the current line is imperative. In talking with the paint equipment manufacturers, the company determined that their current ovens can still be used and ad-

Table 8.5 Concept comparison matrix of current safe line [Reprinted with permission from M. Sondeen].

	Quality			Environment					Cost					
Concepts	Aethetics (Finish)	Security (UL rating)	Fire Proof (UL rating)	Air Waste (mass)	Solid Waste (mass)	Liquid Waste (mass)	Waste Hazard	Human Health Concern	Waste Disposal Cost	Raw Material Cost	Capital Equipment Cost	Labor Cost	Satisfaction	% Difference
VOC Based Paint	9	7	5	5	3	1	3	1	5	5	5	1	518	
Powder Coat	7	7	5	9	7	9	9	7	9	7	3	7	672	30
Current Wash	5	5	5	5	7	1	3	5	7	3	9	5	554	
Recycled Wash	5	5	5	5	5	7	7	5	9	7	5	3	572	3.3
Weights	0.4			0.2					0.4					
Weight (1-10)	9	5	6	7	5	7	7	9	7	3	7	3		
Normalized Weight	45	25	30	20	14	20	20	26	35	15	35	15		
Overall Weight	18	10	12	4	2.9	4	4	5.1	14	6	14	6		

justed to accommodate the powder coating. The same wash system is needed for the preparation of the safe for powder coating so the wash station can also remain unchanged.

This example integrated the DFE tools into the concurrent engineering methodology. As we have seen, small companies face many challenges in implementing industrial ecology principles. However, with the framework provided, the development teams should be able to logically approach the integration of DFE into the development of new products and processes.

8.3 SUMMARY

In this chapter, we illustrated how the principles of industrial ecology can be integrated into the concurrent engineering methodology using DFE tools. Appropriate DFE tools were identified for each design phase, and shown how they integrate into that phase. It is often difficult for small companies to understand and incorporate industrial ecology into their day-to-day operations. With a tool such as the sphere of influence, the company can define the boundaries within which they can affect the environmental performance of their products, processes and operations. Furthermore, as small companies must be cost conscious, even more so than large corporations, quality and cost considerations were considered in the concept comparisons. The example included in this chapter should help convince companies of the feasibility of this approach and help them understand how to apply the DFE tools.

REFERENCES AND BIBLIOGRAPHY

Allenby, B. R., "Industrial Ecology Gets Down to Earth," *Circuits and Devices*, Vol. 10, No. 9, 1994, pp. 24-28.

Carlson-Skalak, S., J. Leschke, M. Sondeen, and P. Gelardi, "E Media's Global Zero: Design for Environment in a Small Company," *Interfaces*, Vol. 30, No. 3, 2000, pp. 66-82.

Frosch, R. and N. Gallopoulos, "Strategies for Manufacturing," *Scientific American*, Vol. 261., No. 3, 1989, pp. 144-152.

Graedel, T. E. and B. R. Allenby, *Industrial Ecology*, AT&T, Prentice Hall, Englwood Cliffs, NJ, 1995.

Graedel, T. E., *Streamlined Life-Cycle Assessment*, Lucent Technologies, Bell Labs Innovations, Prentice Hall, Englewood Cliffs, NJ, 1998.

Rose, C. M., K.A. Beiter, K. Ishii, K. Masui, "Characterization of Product End-of-Life Strategies to Enhance Recyclability," *Proceedings of the ASME Design Engineering Technical Conferences*, Atlanta, paper number DETC98/DTM-5742 on CD-ROM, 1998.

Sondeen, Mark, *Integrating Industrial Ecology into a Concurrent Engineering Framework*, Masters Thesis, University of Virginia, 1999.

Zhang, Y. H., P. Wang, C. Zhang, "Life Cycle Design with Green QFD-II," *Proceedings of the ASME Design Engineering Technical Conferences*, Atlanta, paper number DETC98/DTM-5719 on CD-ROM, 1998.

SUGGESTED READING

Industrial Ecology by T. E. Graedel and B. R. Allenby, Prentice Hall, Englewood Cliffs, NJ, 1995: A good introduction to the concepts of industrial ecology with case studies included.

Streamlined Life-Cycle Assessment by T. E. Graedel, Prentice Hall, Englewood Cliffs, NJ, 1998: Streamlining techniques for the use of LCA framework for use with products, processes, facilities, services, infrastructures, and corporations.

Green Technology and Design for the Environment by Samir Billatoas and Nadia Basaly, Taylor and Francis, Washington, D.C., 1997: Introduction to DFE tools such as design for assembly and disassembly, recycling. Also includes information on regulations, tools for continuous improvement. Includes several case studies.

9

Implementing Concurrent Engineering

In collaboration with Hans-Peter Kemser

Now that we have examined the concurrent engineering methodology in its entirety, the next step is to implement this methodology within your company. Implementing a new methodology like concurrent engineering that encompasses so much of a company's resources cannot occur overnight. The best approach is to start with a pilot project, and then build on that project's success by expanding the implementation with time and more projects. With each new project, the previous implementation should be assessed, improvements made, new tools and technologies incorporated, and new personnel involved. Eventually, the principles of CE will be implemented throughout the company. A series of steps has been developed as a suggested guideline for implementing CE and is shown in Figure 9.1. Following these steps increases the company's chances of successfully implementing CE. Skipping steps may seem to speed the implementation process, but in reality can lead to problems that could lead to unsatisfactory results in the end. The first pilot project should be carefully selected. Employees will be watching for the outcome of the project. The first CE development teams should consist of motivated employees with a commitment to CE, and management needs to take an active interest in the project and its outcome, and give this project their full support. Next, the five implementation steps, identify need for change, preparation, pilot project, pilot project review, and implementation expansion, will be described in more detail.

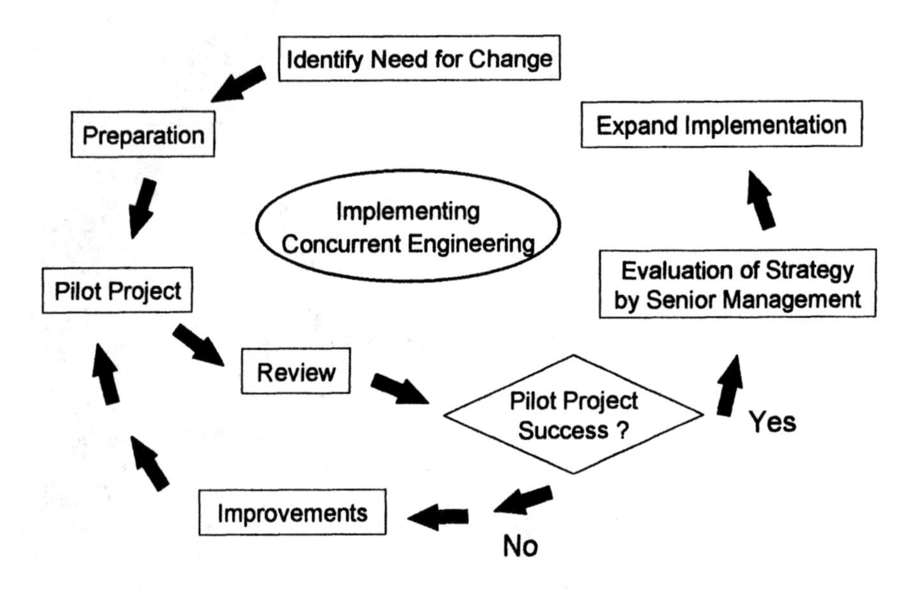

Figure 9.1 Overview of implementation strategy.

9.1 IDENTIFY NEED FOR CHANGE

The company must first recognize a need for changing their current development practices. In most cases, a company is faced by a crisis such as a loss of market share, or an influx of competitive products from overseas that are slowly starting to appear at customer sites. The company may also be facing rising costs that cannot be passed on to the customer due to competitive products, and they must look to product and process improvements that can help them gain a cost advantage. Once the company has recognized the need for changing their development practices and made a commitment to committing resources to implement a change, then they must begin the preparations for incorporating the changes within the company.

9.2 PREPARATION

There are three teams that need to be formed and prepared before the concurrent engineering pilot project can begin: senior management, the project management team, and the cross-functional product and/or process development team. Each of these groups will be discussed separately.

9.2.1 Senior Management

The most important step in the implementation of an operational change like the incorporation of concurrent engineering into a company is the recognition by senior management to change their existing approach. As discussed above, this change can be driven by loss of market share, slowing of revenue growth or even loss of revenue, or competitive pressures. The senior management must recognize the need for change, commit to the change, and support its implementation.

Once management has committed to this undertaking, they must then educate themselves on the concurrent engineering methodology. Additionally, they should seek out other companies that have successfully implemented concurrent engineering to observe their operations. Next, they must analyze their current product and process development methods and compare them to their competitors. It is important that senior management understand concurrent engineering and the difference it can make in their operations.

After educating themselves and collecting as much information as possible, senior management should develop a vision statement that indicates the direction in which the organization will now be headed. It should be easy to communicate and easily understood by all employees. The vision should be communicated to employees throughout the company, and management must show through their actions that they are committed to this change.

Next, senior management must clearly define measurements for the success of the pilot project. These measurements should be carefully considered and should indicate if, where, and how concurrent engineering development is an improvement over their current methods. The range of measurements should cover design performance measurements such as number of design changes required after the conceptual design phase, and product performance measurements such as number of engineering changes after release to production and number of quality problems. Finally, they must select a management team that will be responsible for overseeing the pilot project.

9.2.2 Project Development Management Team

The project development management team (MT) begins its work with the selection of an appropriate pilot project. The following considerations should be taken into account when selecting this project [Smith, 1991]:

- The primary objectives of the first project are to train all personnel and to be successful.
- The project should have a short to moderate time scale, i.e., an evolutionary or incremental design change.

- The project should be large enough to be meaningful, but the project should not undertake risky development tasks that could easily fail under the old system of development.
- Choose a project before the product specifications are defined.

The next task for the MT is the selection of a cross-functional product development team or a cross-functional process development team, whichever is appropriate to the project. In general, the team should include of the following members:

- *Team leader*: in charge of managing the team and project; has a major influence on the success of the project
- *Core team members*: enthusiastic employees of the company with good team and technical skills
- *Facilitator*: this person is different from the team leader; should be well versed in concurrent engineering methodology to help facilitate the teams use of the methodology and help overcome obstacles in adopting the new development methods, this person may also be needed to help facilitate team performance
- *Marketing and sales*: this person is able to give perspective on customer needs, and competitive markets
- *Computer support*: this person is available to help with project documentation and communication.

After the team has been selected, a training plan is developed to educate the team members on the principles of CE, the CE methodology as shown in this book, CE tools and techniques, many of which have been shown in this book, and finally teamwork training. Teamwork training is essential for the success of concurrent engineering. Ideally, it should be performed by someone professionally trained in team formation and facilitation, and they should be available for the duration of the project. There should be an initial class in team performance, but there should also be short refresher courses throughout the project.

Next, the MT needs to develop a performance plan for each team member that emphasizes their individual responsibilities as well as their team responsibilities. The MT needs to work with senior management to put a reward structure in place that rewards and recognizes the team members for their contributions to the project, and most importantly to recognize the success of the team. In addition to explicitly stating expected performance goals, the MT needs to ensure success by encouraging open and frequent communication between team members. This can occur through the use of co-location, in which the team members are moved out of their current departments, and located together for easy access to each other. Co-location shows a commitment to the team and the success of the project.

9.2.3 Cross-Functional Development Teams

The cross-functional development team, whether it is a product or process team, begins its work with the training program. They need to thoroughly understand the principles and implementation of concurrent engineering and how to effectively function as a high performance team. The most effective way to accomplish this training is to have several days of training at the beginning with refreshers and additional information given throughout the duration of the project. Giving all the needed information at the beginning of the project can be overwhelming, and much of it can be forgotten before it is needed.

During the training program, the communication methods within the teams and how they communicate with management should be defined. The team needs to decide how they will distribute information, schedule meetings, distribute results from meetings, and store and receive data. The team also needs to decide how they will implement the documentation method, that is, how will they store and share the outputs from each design step.

Before beginning the pilot project, the team should analyze their current process. Senior management should share what they have already gathered on this subject, but the development team will be able to provide much more detail about the process within which they have been working. The team should assess the following:

- Organizational structure
- Physical locations of different departments involved in the development and support of new products and processes
- Relationship with suppliers
- Current product and process development methods and their associated costs
- Existing tools and techniques
- Customer satisfaction

Collecting this information on the current methods may be very time consuming, since most small companies do not document their design processes. However, analyzing the current methods will help the team identify major areas of improvement for the first project. In addition, this information can be used later to measure how successful the change to concurrent engineering was.

Next, the team must develop a set of target goals for this new undertaking. They should start with the goals defined by senior management and use the competitive data that senior management collected. Setting their own goals and targets helps to empower the team, which is the mark of a good concurrent engineering development team. Finally, the development team needs to develop an implementation strategy which will be developed through the first phase of development, the project planning phase, as covered in Chapters 4 and 5.

9.3 PILOT PROJECT

Now that the teams have been selected, training completed, goals and targets set, the development team can now start the pilot project. The team will follow the phases and design steps outlined in Chapters 4, 5, and 6. Because of their unfamiliarity with the process, and the probable lack of documentation from past projects, the team will probably require more time than if they had a familiarity with the process. Therefore, this project should be seen as a training ground and not to compare development times with the old methods or set guidelines for future projects. The success should be measured more in the product measurements discussed in the section on senior management.

9.4 PILOT PROJECT REVIEW

At the completion of the pilot project, the project development management team and the cross-functional development team should review and evaluate the project. The review should documented and the team should evaluate the following elements:

- Achieved results in comparison with set targets,
- The product and process development methodology,
- The development tools and techniques used and their effectiveness, such as the House of Quality,
- Lessons learned and areas for improvement, and
- Team performance.

Once this review is complete, senior management should review this information with the project management team, and with the project development team. Recall that this is the first step to the implementation of concurrent engineering throughout the company, which is a process that could take a year or more.

9.5 IMPLEMENTATION EXPANSION

At the completion of a successful pilot project, senior management and the project management team need to develop a strategy for expanding the implementation of the concurrent engineering methodology throughout the company. One way this implementation could be achieved is to start with an evaluation of the company's organizational structure to examine where there are large gains to be made in supplier relations, customer satisfaction, and quality or cost savings. These areas should be targeted as the next expansion projects. The organizational structure may need to change to support CE implementation, as discussed in Chapter 2. Also, communication systems need to be examined. Often, even in small companies, different functional organizations have different methods of communications, making communication difficult across functions such as distribution and procurement, if not impossible. These communication barriers should be removed, allowing all employees to communicate with each other. Next, training programs should be put in place to train employees in concurrent engineering methodology and in teamwork. Employees should be empowered to make decisions, so that all decisions are not made at a management level. The reward system should be changed to motivate employees to contribute to the success of concurrent engineering and to reward them for their participation in teams. The incorporation of concurrent engineering throughout the company is a cultural change, which is often unsettling as it requires a behavioral change. This change is more easily accomplished when employees see that senior management is committed to the change and will reward them for work done well under the new system.

9.6 MANAGING THE NEW APPROACH

Management needs to communicate the successes of the new approach to all employees. They need to show how the changes have improved the company's performance and how employees can benefit from the new system. For successful implementation company-wide, the employees should be encouraged to improve the existing processes through a commitment to continuous improvement, one of the cornerstones of the CE methodology. Employees drive the success of CE, and they should be recognized and rewarded for their efforts.

9.7 THE BENEFITS

Once this concurrent engineering methodology has been implemented, your company should begin to see the benefits that were discussed in Chapter 1. First, the problems once recognized as signals for change should be disappearing. Engineers should be spending less time solving customer problems, and have the time to concentrate on new projects. You should see improved customer and supplier relations as they are more involved in the development process. The quality of products should be improved as seen through a reduction of field reported problems, higher yields on the production lines, fewer rework machines, and less scrap. The product and process designs should be improved due to communication among team members and through documentation of the design process and its outputs. These improvements should be seen through a reduction of engineering design changes and process changes. There should be higher employee satisfaction due to an increased sense of ownership in the products and through their empowerment to make decisions within a team concerning their designs. Finally, there should be improvements in getting products to market in a shorter time, a reduction in costs, increased profitability, improved competitiveness, and finally, higher customer satisfaction. You will not see all these improvements right away, but with time and with the experience of using concurrent engineering, the company should reap all these benefits.

REFERENCES AND BIBLIOGRAPHY

Kemser, H., *Concurrent Engineering Applied to Product Development in Small Companies*, Masters Thesis, University of Virginia, 1997.Pahl, G. and W. Beitz, *Engineering Design*, Springer-Verlag, New York, 1996.

Smith, P. and D. Reinertsen, *Developing Products in Half the Time*, Van Nostrand Reinhold, New York, 1991.

Appendix A

House of Quality

The house of quality (HOQ) shown in Figure A1 was developed from interviews conducted by students in an undergraduate design class that was taught at the University of Virginia. The students developed a list of design features and attributes for an ideal ice cream scoop. Then, they interviewed home users of ice cream scoops and had them test four different scoop designs. The students also interviewed employees of Baskins-Robbins who tested the same four scoops. During testing, the students collected information on the likes and dislikes of the scoops and their design features. Next, the interviewees answered a series of questions on what they would like to see in the ideal ice cream scoop. From these interviews, the thirteen customer attributes were rated for importance on a scale of 1 to 5, where 1 is the least important and 5 the most. This information is shown in the left two columns of the HOQ in Figure A1. The far right hand column shows a graphical representation of the customer ratings for each of the attributes of the four designs. The scoop that is to be redesigned is represented by the solid triangle, and the competitive scoops are represented by a plus sign, a circle, and a square. A line is used to connect the solid triangles.

Next, the customer attributes were translated into engineering characteristics shown in the columns. The customer attributes are generally short paraphrases of what the customers have voiced about the product. An engineering requirement is a translation of a customer attribute into a measurable statement. For example, the customer attribute, cuts through ice cream easily, is translated into two requirements: 1) scoop hardness and 2) handle and shank length. In general, there should only be one or two engineering requirements for each customer attribute.

After the engineering characteristics have been defined, the desired trends are noted. A desired trend, shown in the row above the engineering requirements, is used to indicate the desired direction for improving the requirement. For example, to improve the ability of the scoop to cut through ice cream easily, the scoop must not bend when scooping hard ice cream. Therefore, the material hardness should be increased if possible. If there is no trend associated with a requirement, the entry is left blank.

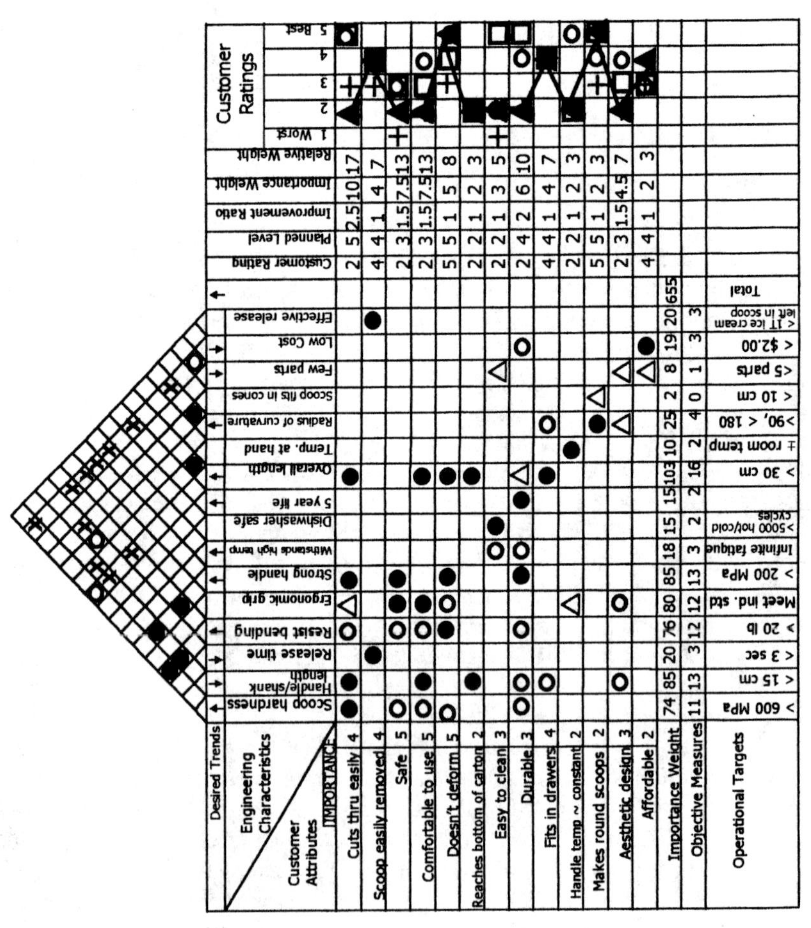

Figure A1 House of quality for an ice cream scoop.

Next, the relationship matrix is completed, which indicates if there is a relationship between a customer attribute and an engineering requirement. If a relationship exists, then a symbol indicating the strength of that relationship is filled in that location. A strong relationship is indicated by a solid circle, a medium relationship by a circle, and a weak one by a triangle. For example, there is a strong relationship between reaching the bottom of the carton and the length of the handle and shank. There is a weak relationship between making round ice cream scoops and fitting into a cone. In this matrix, every row should have at least one entry indicating that every customer attribute has been addressed. However, not every column requires an entry. For example, a regulatory or safety engineering requirement may not be mentioned by the customer, or a technological improvement may not be anticipated.

Now the roof must be completed. The roof shows the relationship between the engineering requirements. There are three types of relationships defined in the matrix: 1) strong positive represented by a solid circle, 2) weak positive represented by a circle, and 3) negative represented by an X. This matrix is used to help the designers see how a change in one requirement impacts others. For example, there is a strongly positive relationship between resist bending and strong handle. If the handle is made stronger then it will also resist bending better. On the other hand, there is a negative relationship between handle length and overall length. The desired trend for the handle length is to decrease it to make scooping hard ice cream easier, however, the desire is to increase the overall length to allow the user to reach the bottom of tall cartons of ice cream. Shortening the handle decreases the overall length, and therefore negatively impacts the desire to increase the overall length of the scoop.

The next task is to define targets for the engineering requirements and the testing methods used to ensure the targets are met. To save room, only the operational targets are listed in Figure A1 at the bottom of the house.

To assist the development team in setting design priorities, importance weights and relative weights for the rows and columns are calculated. We begin with the rows. First, look at the customer rating for the product and the competitive product ratings. The development team needs to decide on a realistic goal for improving each customer attribute. For example, the attribute, cuts through ice cream easily, was rated as a 2 for our scoop, and the other scoops were rated as a 3 and 5s. The team has decided to compete effectively, the new design should have planned level of 5. Not every attribute can be improved to or maintained as a 5, so the team must decide which attributes are the most important and what levels of performance can be realistically achieved. Next, the improvement ratio is calculated by dividing the planned level for a given customer attribute by its customer rating. Next, the importance weight is calculated by multiplying the customer importance by the improvement ratio. Finally, the relative weights for the rows are calculated as follows:

$$\frac{(\text{Importance weight})_i}{\sum_{i=1}^{r}(\text{Importance weight})} * 100 \qquad \text{(Equation A1)}$$

where r is the number of customer attributes.

A similar process is then carried out for the columns. First, a column's importance weight is calculated by multiplying the customer importance by the strength of the relationship. A solid circle is worth 5 points, a circle is worth 3 points, and a triangle is worth 1 point. For example, in the first column, there is one solid circle and the rest are circles. The importance weight for this column is calculated as follows:

$$4(5)+5(3)+5(3)+5(3)+3(3)=74 \qquad \text{(Equation A2)}$$

Next, the relative weight for each column is calculated as:

$$\frac{(\text{Importance weight})_i}{\sum_{i=1}^{c}(\text{Importance weight})} * 100 \qquad \text{(Equation A3)}$$

where c is the number of engineering characteristics.

Now that the house of quality is finished, the team can use the house to help them establish design priorities. For example, the most important customer attribute is 'cuts through ice cream easily', followed by safety and comfort. In the engineering characteristics, the overall length, strength, and handle and shank length are the most important followed closely by bending resistance and ergonomic grip. The least important attributes are reaching the bottom of the carton, handle temperature, and scoop roundness. The technical characteristics that are the least important are few parts, temperature at hand, long life, withstands high temperature, and dishwasher safe. The design priorities can now easily be established by the design team, and using the roof, they can understand the trade-offs that occur as engineering characteristics are changed.

Appendix B

Risk Assessment Tool for Sheetmetal and Plastic Parts[1]

This risk assessment tool for sheetmetal and plastic parts should be used in conjunction with a product development plan. This tool contains a series of questions that lead the team members through the process of assessing risk. This tool is comprised of two types of questions. The product development questions are meant only to get the team members thinking about the details of the part that is being developed, organize their thoughts, and serve as a way to document their thoughts at the time of risk assessment. The risk questions give the team members a chance to express in qualitative terms their assessment of risk in different areas.

The team members are asked to answer each question to the best of their knowledge. The answers to the questions should be entered in the 'answer' column. In the case of a risk assessment question, the team members should enter the number value of the answer that best represents their opinion. The numerical answers to all the risk assessment questions are recorded in the risk assessment charts. These charts provide the team members a quick reference to see what their assessment of the risk is in different areas.

Comment boxes are provided with each question to allow the team members to note important information about the development project. Often, these comments can be instrumental in understanding why a decision was made when reviewing the information. Finally, the team should document any lessons learned at the end of the risk assessment process. This documentation encourages the team to review the design process and to share their knowledge with others.

This risk assessment tool does not completely address all risk or contingencies. If the team members feel that something is not included in this tool, they are certainly encouraged to add this information.

---

[1] The material in this appendix is reprinted with permission from M. Born and comes from the following work: Born, M., *The Assessment of Risk in the Development of Sheetmetal and Plastic Parts*, Masters Thesis, University of Virginia, May 2000, pp. 72-100.

Table B1. General questions for the risk assessment tool.

Part I: GENERAL QUESTIONS		
Questions	**Time/Cost**	**Comments**
1. What is the expected time frame for the development of this part?		
2. When will functional parts be needed for testing?		
3. What is the estimated cost of the part?		

Table B2. Level one design questions for the risk assessment tool.

Level One Questions			
Questions	**Options**	**Answers**	**Comments**
1. Describe what the function this component will perform.			
2. How well do you understand the function you just described?	a. Very well (0 pts) b. Well (1 pt) c. Somewhat well (2 pts) d. Undecided (3 pts) e. Do not understand (4 pts)		
3. What type of design is it?	a. Incremental (0 pts) b. Evolutionary (1 pt) c. Original (3 pts) d. Breakthrough (4 pts)		
4. Should the part be one piece or an assembly?			
5. Will this product be standalone?	a. Yes b. No Note: If yes, then go to question #6.		
5a. Where within the final product is the part located?			
5b. Explain how it interacts with mating parts.			
5c. With these interactions in mind, how stable do you expect the system to be?	a. Very stable (0 pts) b. Stable (1 pt) c. Somewhat stable (2 pts) d. Not sure (3 pts) e. Not stable (4 pts)		
6. How durable are the parts expected to be? (This will affect material selection)			
7. Will you use sheetmetal or plastic?			

Table B3. Level one manufacturing questions for the risk assessment tool.

Level One Questions			
Questions	**Options**	**Answers**	**Comments**
8. What volumes (in parts per unit time) are you expecting to produce?.			
9. Taking into consideration the function, material, and volume, what type of process(es) are needed to manufacture the part?			
10. Is the technology for this process available?	a. Yes [0 pts] b. No [4 pts]		
11. What is your experience level with the technology required to manufacture this part?	a. Extensive [0 pts] b. Some [1 pt] c. Very little [2 pts] d. Not sure [3 pts] e. None [4 pts]		
12. Do you have the resources needed to manufacture the part in-house or will it be out-sourced?	a. If out-sourced, proceed to question number 13. b. If in-house, proceed to question number 14.		
13. In the manufacturing supplier matrix, please list the possible suppliers.			
14. Rate the suppliers in the manufacturing supplier matrix.			

Table B4. Level one tooling questions for the risk assessment tool.

Level One Questions			
Questions	Options	Answers	Comments
15. What is the tooling budget?			
16. Are the resources available to successfully develop the tool in-house or will it be out-sourced?	If out-sourced, proceed to question number 17. If in-house, proceed to question number 18.		
17. In the tooling supplier matrix, list all possible suppliers.			
18. In the tooling supplier matrix, rate each supplier.			

Table B5. Manufacturing supplier matrix for the risk assessment tool.

	Manufacturing Supplier Matrix						
	Name	Necessary Personnel	Available Equipment	Necessary Technology	Quality Monitoring (Cp, Cpk)	Financial Stability	Material Sourcing
1.							
2.							
3.							
4.							
5.							
6.							
7.							
8.							
9.							
10.							

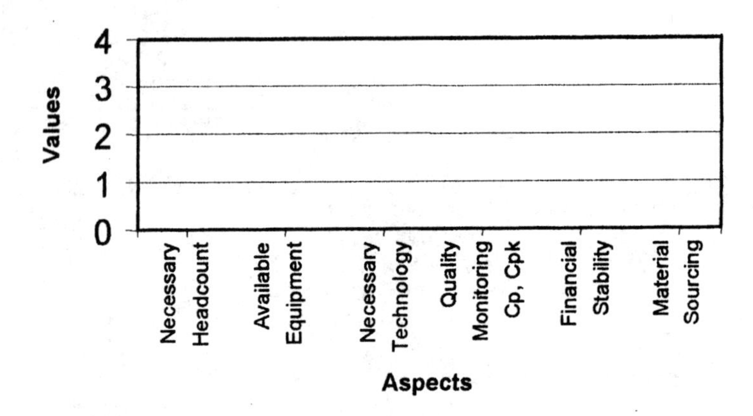

Figure B1. Manufacturing supplier graph for representation of supplier aspects.

Table B6. Tooling supplier matrix for the risk assessment tool.

	Tooling Supplier Matrix							
	Name	Necessary Personnel	Available Equipment	Necessary Technology	Quality Monitoring (Cp, Cpk)	Financial Stability	Material Sourcing	Post-Delivery Support
1.								
2.								
3.								
4.								
5.								
6.								
7.								
8.								
9.								
10.								

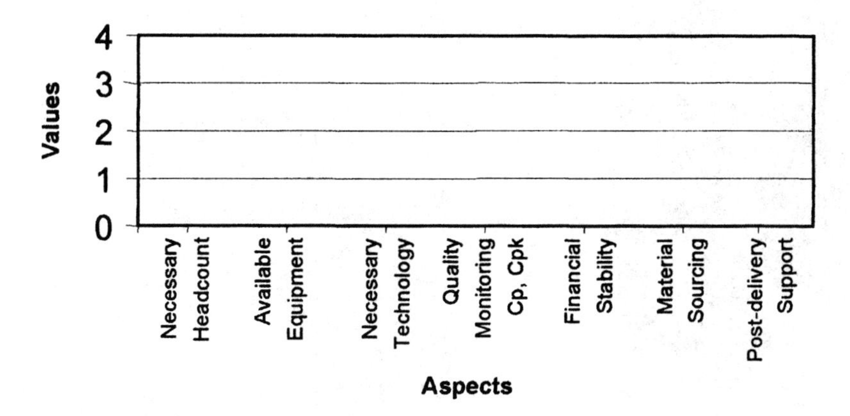

Figure B2. Tooling supplier graph for representation of supplier aspects.

Table B7. Level two design questions for sheetmetal parts.

Level Two Questions: Sheetmetal			
Questions	**Options**	**Answers**	**Comments**
1. Determine the shape and dimensions of the part.			
2. Identify the critical part dimensions.			
3. Does this part require special insulation properties? If so, describe them.			
4. What type of sheetmetal should be used? Should it be corrosive or erosive?			
5. What are the cosmetic requirements for the part?			
6. Given these cosmetic requirements, should a pre-plated material or a post-plated material be used?			
7. If you need to join sheetmetal parts or corners, rate the following aspects in terms of needed functionality/performance.	a. Very effective (0 pts) b. Effective (1 pt) c. Somewhat effective (2 pts) d. Not sure (3 pts) e. Not effective (4 pts)		
7a. Tog-L-Loc			
7b. Extrude and roll			
7c. Rivet			
7d. Spot weld			
7e. Tabs			
8. Considering the chosen material, the design, and the process, how confident are you that your durability expectation will be met?	a. Very confident (0 pts) b. Confident (1 pt) c. Somewhat confident (2 pts) d. Not sure (3 pts) e. Not confident (4 pts)		

Table B8. Level two design questions for sheetmetal parts continued.

Level Two Questions: Sheetmetal			
Questions	**Options**	**Answers**	**Comments**
9. Will the part require assembly, and if so, what type?			
9a. Taking assembly methods into consideration, how compatible is this design with the method of assembly?	a. Very compatible [0 pts] b. Compatible [1 pt] c. Somewhat compatible [2 pts] d. Not sure [3 pts] e. Not compatible [4 pts]		
10. Have you considered the issues related to the maintenance of the part once it is in use?			
10a. How easy will maintenance be to perform on this design?	a. Very easy [0 pts] b. Easy [1 pt] c. Somewhat easy [2 pts] d. Not sure [3 pts] e. Difficult [4 pts]		
11. Has the part been designed with recycleability of materials in mind?			
11a. How easy will it be to recycle this part?	a. Very easy [0 pts] b. Easy [1 pt] c. Somewhat easy [2 pts] d. Not sure [3 pts] e. Difficult [4 pts		
12. Has the part been designed with end-of-life disassembly in mind?			
12a. How easy will it be to disassemble this part?	a. Very easy [0 pts] b. Easy [1 pt] c. Somewhat easy [2 pts] d. Not sure [3 pts] e. Difficult [4 pts]		

Table B9. Level two manufacturing questions for sheetmetal parts.

Level Two Questions: Sheetmetal			
Questions	Options	Answers	Comments
13. How confident are you that the parts can be produced with the specified features?	a. Very confident [0 pts] b. Confident [1 pt] c. Somewhat confident [2 pts] d. Not sure [3 pts] e. Not confident [4 pts]		
14. Re-evaluate the preliminary process selection to ensure that it is optimal. If the process has changed, please specify which process will be used.			
15. How confident are you that you will be able to manufacture the parts with the specified cosmetic requirements?	a. Very confident [0 pts] b. Confident [1 pt] c. Somewhat confident [2 pts] d. Not sure [3 pts] e. Not confident [4 pts]		
16. How easy is the part to manufacture?	a. Very easy [0 pts] b. Easy [1 pt] c. Somewhat easy [2 pts] d. Not sure [3 pts] e. Difficult [4 pts]		
16a. If question 16 scored a value of 2 or higher, what are some of the problems you anticipate in manufacturing this part?			
16b. If question 16 scored a value of 2 or higher, how feasible is it to improve the design or change the process and still remain with in the development schedule?	a. Very feasible [0 pts] b. Somewhat feasible [1 pt] c. Marginally feasible [2 pts] d. Not sure [3 pts] e. Not feasible [4 pts]		

Table B10. Level two manufacturing questions for sheetmetal parts continued.

Level Two Questions: Sheetmetal			
Questions	**Options**	**Answers**	**Comments**
17. Is capability to measure the critical features specified in question 2 of the "design" section available?	a. Yes [0 pts] b. No [4 pts]		
17a. Do you expect to achieve operator consistency?	a. Yes [0 pts] b. Not sure [2 pts] c. No [4 pts]		
18. Will you build an inventory or use a form of just-in-time delivery or some other approach? Please specify.			
19. Will the part be dual sourced?			
19a. If dual-sourced, how successful do you expect to be in qualifying both suppliers in terms of quality?	a. Very successful [0 pts] b. Successful [1 pt] c. Somewhat successful [2 pts] d. Not sure [3 pts] e. Unsuccesful [4 pts]		
19b. How confident are you that the parts will be compatible/interchangeable?	a. Very confident [0 pts] b. Confident [1 pt] c. Somewhat confident [2 pts] d. Not sure [3 pts] e. Not confident [4 pts]		
20. If the part requires assembly or is part of an assembly, will you:			
20a. Assemble at the supplier?	Yes No		
20b. Assemble at an outside location (i.e., a distribution center)?	Yes No		
20c. Other option? Please specifiy.			

Table B11. Level two manufacturing questions for sheetmetal parts continued.

Level Two Questions: Sheetmetal			
Questions	**Options**	**Answers**	**Comments**
21. How successful do you think you will be at:			
21a. Avoiding possible delays due to shipping parts from two different suppliers?	a. Very successful [0 pts] b. Successful [1 pt] c. Somewhat successful [2 pts] d. Not sure [3 pts] e. Unsuccesful [4 pts]		
21b. Coordinating the assembly effort when parts arrive from two different suppliers?	a. Very successful [0 pts] b. Successful [1 pt] c. Somewhat successful [2 pts] d. Not sure [3 pts] e. Unsuccesful [4 pts]		
22. If the part requires assembly or is part of an assembly, what special alignment features are needed?			
22a. How will the parts be assembled (operator, machinery, etc.) Please specify.			
22b. How easy is the assembly process?	a. Very easy [0 pts] b. Easy [1 pt] c. Somewhat easy [2 pts] d. Not sure [3 pts] e. Difficult [4 pts]		
22c. How sensitive is the part to vibrations and impact due to shipping?	a. Not sensitive [0 pts] b. Not very sensitive [1 pt] c. Sensitive[2 pts] d. Very sensitive [3 pts] e. Not sure [4 pts]		

Table B12. Level two manufacturing questions for sheetmetal parts continued.

Level Two Questions: Sheetmetal			
Questions	**Options**	**Answers**	**Comments**
22d. Taking the answer to question 22c into consideration, how –will the parts be shipped so they can be assembled later?			
22e. How will the parts be shipped to their final destination (if shipped assembled) or from the supplier(s) to the assembly location (if parts are shipped separately)?			
22f. Has a drop test been performed on the part?	Yes No		
22g. How did the part perform in the drop test?	a. Very well [0 pts] b. Well [1 pt] c. Marginally [2 pts] d. Not sure [3 pts] e. Poorly [4 pts]		
22h. Given the answer to question 22g, will you have to make any changes either to the design or to its shipping method to improve the drop test performance?	Yes No		
23. Are there any concerns regarding the operating environment of the part?			
23a. What types of environmental test need to be performed to address concerns?			
23b. How well did the part perform these tests?	a. Very well [0 pts] b. Well [1 pt] c. Marginally [2 pts] d. Not sure [3 pts] e. Poorly [4 pts]		

Table B13. Level two tooling questions for sheetmetal parts.

Level Two Questions: Sheetmetal			
Questions	**Options**	**Answers**	**Comments**
24. What type of die (progressive, individual, no die – lasercutting) will be used?			
25. What is the lead time for the tool?			
26. Do you need to build a prototype tool to test the critical features?	Yes No		
26a. If testing of the critical features is needed, can you build the tool in modules and work concurrently on building the tool while testing the critical features?	a. Yes [0 pts] b. Not sure [3 pts] c. No [4 pts] Note: If unable to build in modules could pose a risk to part development because tool lead-times may not be met.		
26b. With this information in mind, will you be able to afford this prototype tool?	a. Yes [0 pts] b. Not sure [3 pts] c. No [4 pts]		
27. Given the tool information, how successful will you be in achieving the needed volumes?	a. Very successful [0 pts] b. Successful [1 pt] c. Somewhat successful [2 pts] d. Not sure [3 pts] e. Unsuccesful [4 pts]		
27a. If question 27 scored a 2 or higher, did you build a prototype tool?	a. Yes [proceed to question 27b] b. No [proceed to question 27c]		
27b. If the answer to question 27a is yes, will you be able to use this prototype tool as a backup tool to achieve the needed volumes?	a. Yes [stop] b. No [proceed to question 27e]		

Table B14. Level two tooling questions for sheetmetal parts continued.

Level Two Questions: Sheetmetal			
Questions	**Options**	**Answers**	**Comments**
27c. If the answer to question 27a is no, how successful would you be at making up the volume lost after having your tool down for the amount of time occurring in a worst-case scenario?	a. Very successful [0 pts] b. Successful [1 pt] c. Somewhat successful [2 pts] d. Not sure [3 pts] e. Unsuccesful [4 pts]		
27d. If the answer to question 27c is 2 or above, you need to order a backup tool.			
27e. Will you be able to afford a backup tool?	a. Yes [0 pts] b. Not sure [3 pts] c. No [4 pts]		

Table B15. Level two design questions for plastic parts.

Level Two Questions: Plastic			
Questions	**Options**	**Answers**	**Comments**
1. Determine the shape and dimensions of the part.			
2. Identify the critical part dimensions.			
3. Does this part require special insulation properties? If so, describe them.			
4. Given the current part design, what type of plastic should be used?			
5. What are the cosmetic requirements for the part?			

Table B16. Level two design questions for plastic parts continued.

Level Two Questions: Plastic			
Questions	**Options**	**Answers**	**Comments**
6. Given these cosmetic requirements, have you located the weld lines and gates?			
7. How much draft do you need to include in your design?			
8. Determine the hygroscopic properties of the material.?			
8a. How will these hygroscopic properties affect the functionality and processing of the part?			
8b. How will the effect of the hygroscopic properties impact the part?	a. Very positively [0 pts] b. Positively [1 pt] c. Somewhat positively [2 pts] d. Not sure [3 pts] e. Negatively [4 pts]		
9. Determine how gate location, part shape, the dimensions, and the viscosity of the chosen plastic affect how the material will fill the mold.			
9a. How acceptable is this filling?	a. Very acceptable [0 pts] b. Acceptable [1 pt] c. Somewhat acceptable [2 pts] d. Not sure [3 pts] e. Unacceptable [4 pts]		

Table B17. Level two design questions for plastic parts continued.

Level Two Questions: Plastic			
Questions	**Options**	**Answers**	**Comments**
10. Determine how gate location, part shape, dimensions, and viscosity will affect flash in the part.			
10a. How acceptable is this flash?	a. Very acceptable [0 pts] b. Acceptable [1 pt] c. Somewhat acceptable [2 pts] d. Not sure [3 pts] e. Unacceptable [4 pts]		
11. Will it be necessary to add fillers to the material to achieve enhanced properties such as lubrication, strength, creep avoidance, flame retardation, or to change electrical or thermal conductivity properties?			
11a. How will these filler affect the cosmetic requirements for the part?			
12. Considering the chosen material, the design, and the process, how confident are you that your durability expectation will be met?	a. Very confident [0 pts] b. Confident [1 pt] c. Somewhat confident [2 pts] d. Not sure [3 pts] e. Not confident [4 pts]		
13. If the part will require assembly, what type of assembly will be required?			

Table B18. Level two design questions for plastic parts continued.

Level Two Questions: Plastic			
Questions	Options	Answers	Comments
13a. Taking into consideration how the part will be assembled, how compatible is this design with the method of assembly?	a. Very compatible [0 pts] b. Compatible [1 pt] c. Somewhat compatible [2 pts] d. Not sure [3 pts] e. Incompatible [4 pts]		
14. Have you considered the issues related to the future maintenance of the part once the customer has it?			
14a. How easy will it be to perform routine maintenance on this design?	a. Very easy [0 pts] b. Easy [1 pt] c. Somewhat easy [2 pts] d. Not sure [3 pts] e. Difficult [4 pts]		
15. Have you designed this part with recyclability of the material(s) in mind?			
15a. How easy will it be to recycle this part?	a. Very easy [0 pts] b. Easy [1 pt] c. Somewhat easy [2 pts] d. Not sure [3 pts] e. Difficult [4 pts]		
16. Have you designed this part with end-of-life disassembly in mind?			
16a. How easy will it be to disassemble this part?	a. Very easy [0 pts] b. Easy [1 pt] c. Somewhat easy [2 pts] d. Not sure [3 pts] e. Difficult [4 pts]		

Table B19. Level two manufacturing questions for plastic parts.

Level Two Questions: Plastic			
Questions	**Options**	**Answers**	**Comments**
17. How confident are you that you will be able to produce parts with the specified features?	a. Very confident [0 pts] b. Confident [1 pt] c. Somewhat confident [2 pts] d. Not sure [3 pts] e. Not confident [4 pts]		
18. Re-evaluate the preliminary process selection to ensure that it is optimal for the part design. If the manufacturing process is changed, please specify which one will be used.			
19. How confident are you that you will be able to manufacture the parts with the specified cosmetic requirements?	a. Very confident [0 pts] b. Confident [1 pt] c. Somewhat confident [2 pts] d. Not sure [3 pts] e. Not confident [4 pts]		
20. How manufacturable is the part?	a. Very manufacturable [0 pts] b. Manufacturable [1 pt] c. Somewhat manufacturable [2 pts] d. Not sure [3 pts] e. Infeasible [4 pts]		
21. Is the necessary capability to measure the critical features specified in question 2 of the design section available?	a. Yes [0 pts] b. No [4 pts]		
21a. Do you expect to achieve consistency?	a. Yes [0 pts] b. Not sure [3 pts] c. No [4 pts]		

Table B20. Level two manufacturing questions for plastic parts continued.

Level Two Questions: Plastic			
Questions	**Options**	**Answers**	**Comments**
22. Will you build an inventory or use a form of just-in-time delivery or another approach? Please specify.			
23. Will the part be dual sourced?			
23a. If dual-sourced, how successful do you expect to be in qualifying both suppliers in terms of quality?	a. Very successful [0 pts] b. Successful [1 pt] c. Somewhat successful [2 pts] d. Not sure [3 pts] e. Unsuccesful [4 pts]		
23b. How confident are you that the parts will be compatible/interchangeable?	a. Very confident [0 pts] b. Confident [1 pt] c. Somewhat confident [2 pts] d. Not sure [3 pts] e. Not confident [4 pts]		
24. If the part requires assembly or is part of an assembly, will you:			
24a. Assemble at the supplier?	Yes No		
24b. Assemble at an outside location (i.e., a distribution center)?	Yes No		
24c. Other option? Please specifiy.			

Table B21. Level two manufacturing questions for plastic parts continued.

Level Two Questions: Plastic			
Questions	**Options**	**Answers**	**Comments**
25. How successful do you think you will be at:			
25a. Avoiding possible delays due to shipping parts from two different suppliers?	a. Very successful [0 pts] b. Successful [1 pt] c. Somewhat successful [2 pts] d. Not sure [3 pts] e. Unsuccesful [4 pts]		
25b. Coordinating the assembly effort when parts arrive from two different suppliers?	a. Very successful [0 pts] b. Successful [1 pt] c. Somewhat successful [2 pts] d. Not sure [3 pts] e. Unsuccesful [4 pts]		
26. If the part requires assembly or is part of an assembly, what special alignment features are needed?			
26a. How will the parts be assembled (operator, machinery, etc.) Please specify.			
26b. How easy is the assembly process?	a. Very easy [0 pts] b. Easy [1 pt] c. Somewhat easy [2 pts] d. Not sure [3 pts] e. Difficult [4 pts]		
26c. How sensitive is the part to vibrations and impact due to shipping?	a. Not sensitive [0 pts] b. Not very sensitive [1 pt] c. Sensitive[2 pts] d. Very sensitive [3 pts] e. Not sure [4 pts]		

Table B22. Level two manufacturing questions for plastic parts continued.

Level Two Questions: Plastic			
Questions	**Options**	**Answers**	**Comments**
26d. Taking the answer to question 26c into consideration, how –will the parts be shipped so they can be assembled later?			
26e. How will the parts be shipped to their final destination (if shipped assembled) or from the supplier(s) to the assembly location (if parts are shipped separately)?			
26f. Has a drop test been performed on the part?	Yes No		
26g. How did the part perform in the drop test?	a. Very well [0 pts] b. Well [1 pt] c. Marginally [2 pts] d. Not sure [3 pts] e. Poorly [4 pts]		
26h. Given the answer to question 26g, will you have to make any changes either to the design or to its shipping method to improve the drop test performance?	Yes No		
27. Are there any concerns regarding the operating environment of the part?			
27a. What types of environmental test need to be performed to address concerns?			
27b. How well did the part perform these tests?	a. Very well [0 pts] b. Well [1 pt] c. Marginally [2 pts] d. Not sure [3 pts] e. Poorly [4 pts]		

Table B23. Level two tooling questions for plastic parts.

Level Two Questions: Plastic			
Questions	**Options**	**Answers**	**Comments**
28. Will the mold have a perpendicular or a stepped parting line?			
29. How many ejector pins are needed and where should they be placed?			
30. How will the addition of any fillers affect the mold life?	a. Very much [4 pts] b. Moderately [2 pt] c. Very little [1 pt] d. Not sure [3 pts] e. Will not affect [0 pts]		
31. What is the lead time for the tool?			
32. Do you need to build a prototype tool to test critical features?	Yes No		
32a. If testing of critical features is needed, can you build the tool in modules and work concurrently on building the tool while testing critical features?	a. Yes [0 pts] b. Not sure [3 pts] c. No [4 pts] Note: Not being able to build the tool in modules in this case could pose great risk to the development of the part because tooling lead-times may cause delay.		
32b. With this information in mind, will you be able to afford this prototype tool?	a. Yes [0 pts] b. Not sure [3 pts] c. No [4 pts]		
33. Given the tool information, how successful will you be in achieving the needed volumes?	a. Very successful [0 pts] b. Successful [1 pt] c. Somewhat successful [2 pts] d. Not sure [3 pts] e. Unsuccessful [4 pts]		

Table B24. Level two tooling questions for plastic parts continued.

Level Two Questions: Sheetmetal			
Questions	**Options**	**Answers**	**Comments**
33a. If question 33 scored a 2 or higher, did you build a proto-type tool?	a. Yes [proceed to question 33b] b. No [proceed to question 33c]		
33b. If the answer to question 33a is yes, will you be able to use this prototype tool as a backup tool to achieve the needed volumes?	a. Yes [stop] b. No [proceed to question 33e]		
33c. If the answer to question 33a is no, how successful would you be at making up the volume lost after having your tool down for the amount of time occurring in a worst-case scenario?	a. Very successful [0 pts] b. Successful [1 pt] c. Somewhat suc-cessful [2 pts] d. Not sure [3 pts] e. Unsuccesful [4 pts]		
33d. If the answer to question 33c is 2 or higher, you need to order one of the fol-lowing: a backup tool, manifold, or backup hot runner manifold.			
33e. Will you be able to afford a backup tool, manifold, or backup hot runner manifold?	a. Yes [0 pts] b. Not sure [3 pts] c. No [4 pts]		

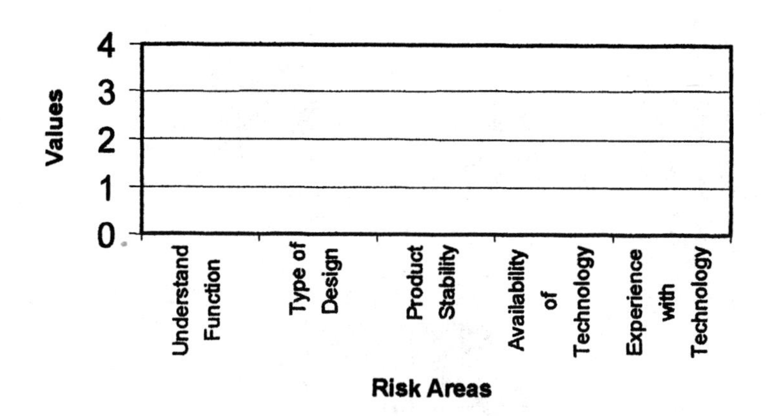

Figure B3. Level one risk assessment graph for sheetmetal parts.

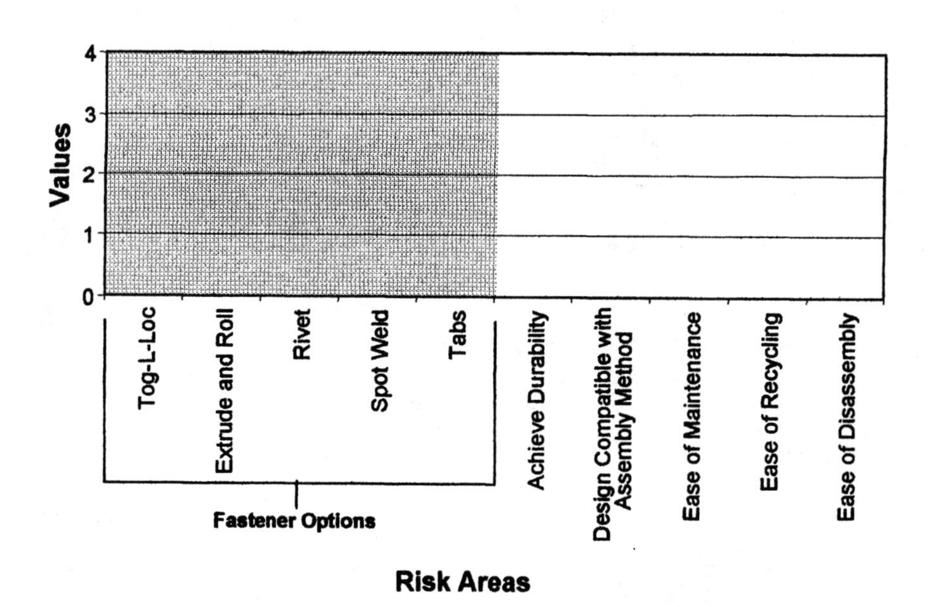

Figure B4. Level two design risk assessment graph for sheetmetal parts.

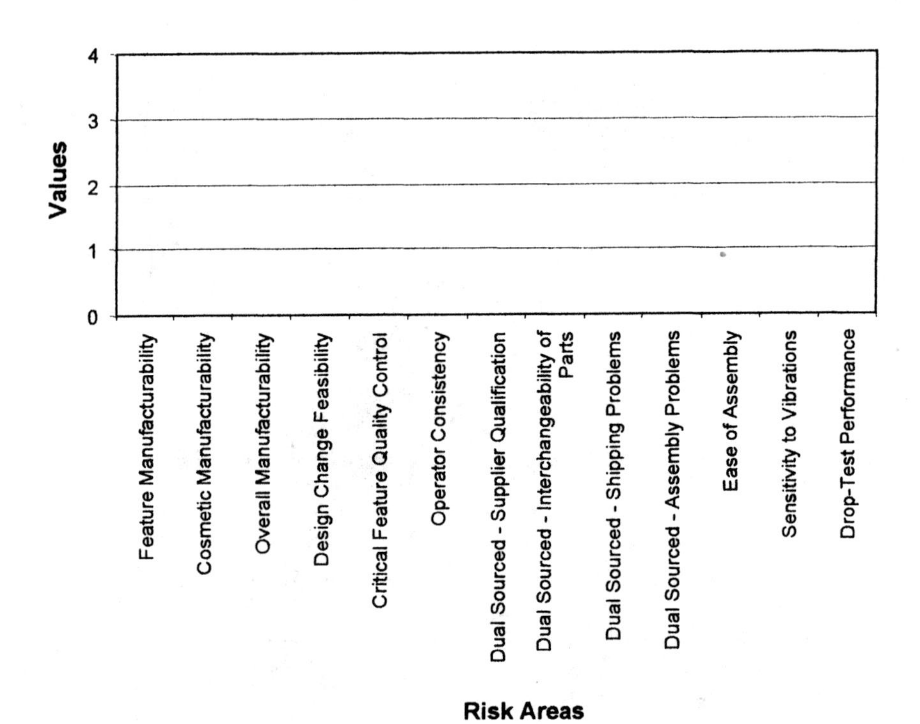

Figure B5. Level two manufacturing risk assessment graph for sheetmetal parts.

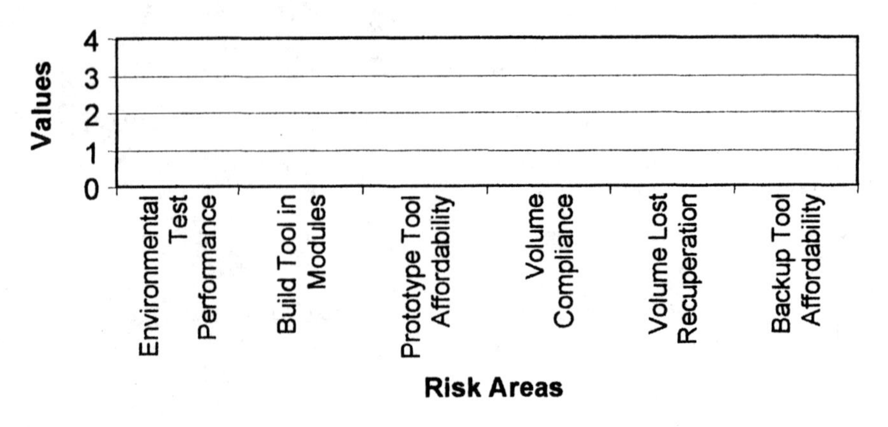

Figure B6. Level two tooling risk assessment graph for sheetmetal parts.

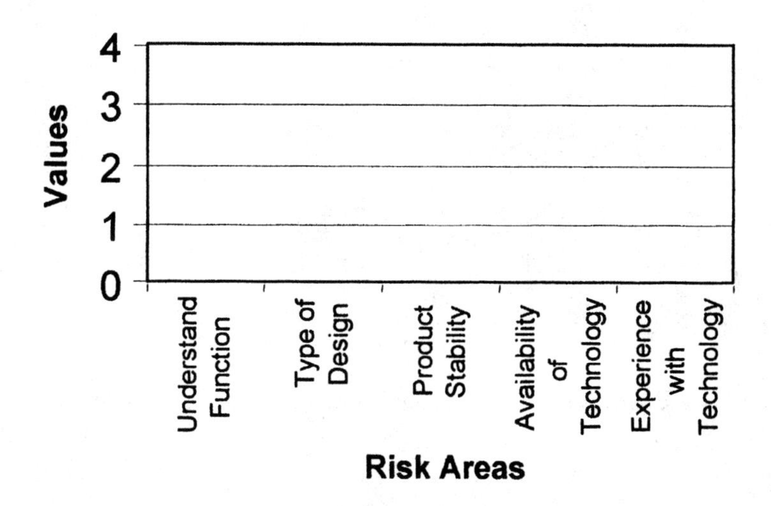

Figure B7. Level one risk assessment graph for plastic parts.

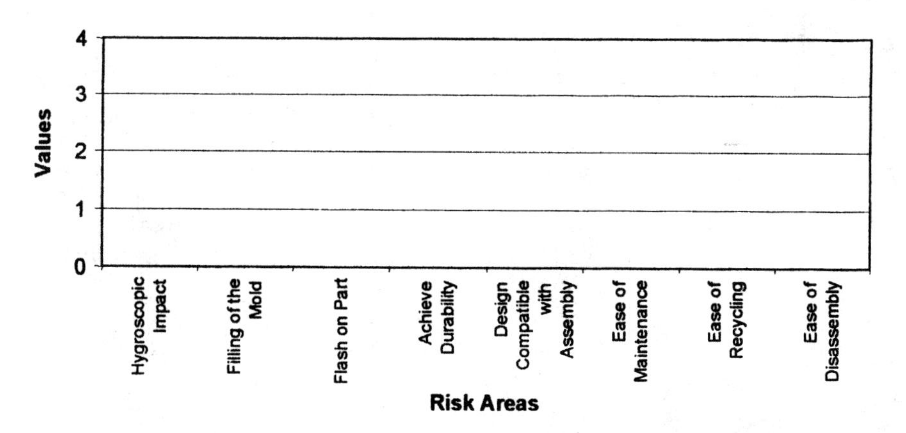

Figure B8. Level two design risk assessment graph for plastic parts.

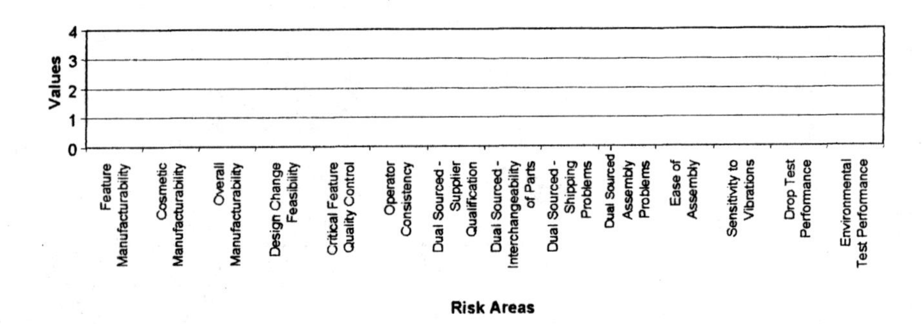

Figure B9. Level two manufacturing risk assessment graph for plastic parts.

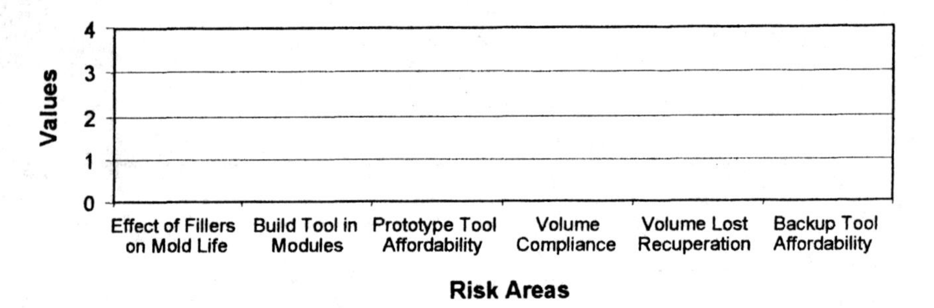

Figure B10. Level two tooling risk assessment graph for plastic parts.

Table B25. Level one questions data table.

Risk Area	Risk Value
Understand function	
Type of design	
Product stability	
Availability of technology	
Experience with technology	

Table B26. Level two questions data table for sheetmetal parts.

Area	Risk Area	Risk Value
Design	Tog-L-Loc	
Design	Extrude and roll	
Design	Rivet	
Design	Spot Weld	
Design	Design compatibility with assembly method	
Design	Ease of maintenance	
Design	Ease of recycling	
Design	Easy of disassembly	
Manufacturing	Feature manufacturability	
Manufacturing	Cosmetic manufacturability	
Manufacturing	Overall manufacturability	
Manufacturing	Design change feasibility	
Manufacturing	Critical feature quality control	
Manufacturing	Operator consistency	
Manufacturing	Dual-sourced supplier qualification	
Manufacturing	Dual-sourced part interchangeability	
Manufacturing	Dual-sourced shipping problems	
Manufacturing	Dual-sourced assembly problems	
Manufacturing	Ease of assembly	
Manufacturing	Sensitivity to vibrations	
Manufacturing	Drop test performance	
Tooling	Environmental performance test	
Tooling	Modularity of tool	
Tooling	Prototype tool affordability	
Tooling	Volume compliance	
Tooling	Lost volume recuperation	
Tooling	Backup tool affordability	

Table B27. Level two questions data table for plastic parts.

Area	Risk Area	Risk Value
Design	Hygroscopic impact	
Design	Filling of the mold	
Design	Flash on part	
Design	Durability achievement	
Design	Design compatibility with assembly method	
Design	Ease of maintenance	
Design	Ease of recycling	
Design	Easy of disassembly	
Manufacturing	Feature manufacturability	
Manufacturing	Cosmetic manufacturability	
Manufacturing	Overall manufacturability	
Manufacturing	Design change feasibility	
Manufacturing	Critical feature quality control	
Manufacturing	Operator consistency	
Manufacturing	Dual-sourced supplier qualification	
Manufacturing	Dual-sourced part interchangeability	
Manufacturing	Dual-sourced shipping problems	
Manufacturing	Dual-sourced assembly problems	
Manufacturing	Ease of assembly	
Manufacturing	Sensitivity to vibrations	
Manufacturing	Drop test performance	
Tooling	Environmental test performance	
Tooling	Effect of fillers on mold life	
Tooling	Modularity of tool	
Tooling	Prototype tool affordability	
Tooling	Volume compliance	
Tooling	Lost volume recuperation	
Tooling	Backup tool affordability	

Appendix C

Streamlined Life Cycle Assessment Matrix Scoring Guidelines[1]

The streamlined life cycle assessment matrix shown in Chapter 8, Table 8.2 is explained more fully in this appendix. For each matrix entry, a score is given based on present conditions. The rating system provided is a guideline and may require adjustment for different types of products.

PRODUCT MATRIX ELEMENT 1, 1

Life Stage: Premanufacture

Environmental Stressor: Materials Choice

If any of the following conditions apply, the matrix element rating is 0:

- For the case where supplier components/subsystems are used: No/little information is known about the chemical content in supplied products and components.
- For the case where materials are acquired from suppliers: a scarce material is used where a reasonable alternative is available. (Scarce materials are defined as antimony, beryllium, boron, cobalt, chromium, gold, mercury, the platinum metals (Pt, Ir, Os, Pa, Rh, Ru], silver, thorium, and uranium.)

If the following condition applies, the matrix element rating is a 4:

- No virgin material is used in incoming components or materials.

If neither of the preceding ratings is assigned, complete the checklist below. Assign a rating of 1, 2, or 3 depending on the degree to which the product meets DFE preferences for this matrix element.

---

[1] The material in this appendix is adapted from Streamlined Life Cycle Assessment by Graedel, © Adapted by permission of Pearson Education, Inc., Upper Saddle River, NJ.

- Is the product designed to minimize the use of materials in restricted supply (see the preceding list)?
- Is the product designed to utilize recycled materials or components wherever possible?

PRODUCT MATRIX ELEMENT 1, 2

Life Stage: Premanufacture

Environmental Stressor: Energy Use

If the following condition applies, the matrix element rating is 0:

- One or more of the principal materials used in the product requires energy-intensive extraction and suitable alternative materials are available that do not. (Materials requiring energy-intensive extraction are defined as virgin aluminum, virgin steel, and virgin petroleum).

If the following condition applies, the matrix element is a 4:

- Negligible energy is needed to extract or ship the materials or components for this product.

If neither of the preceding ratings is assigned, complete the checklist below. Assign a rating of 1, 2, or 3 depending on the degree to which the product meets DFE preferences for this matrix element.

- Is the product designed to minimize the use of virgin materials whose extraction is energy-intensive?
- Does the product design avoid or minimize the use of virgin materials whose transport to and from the facility will require significant energy use? (Such materials are defined as those with a specific gravity preceding 7.0.)
- Is transport distance of incoming materials and components minimized?

PRODUCT MATRIX ELEMENT 1, 3

Life Stage: Premanufacture

Environmental Stressor: Solid Residues

If either of the following conditions apply, the matrix element rating is 0:

- For the case where materials are acquired from suppliers: Metals from virgin ores are used, creating substantial waste rock residues that could be avoided by the use of recycled material, and suitable recycled material is available from recycling streams.
- For the case where supplier components/subsystems are used: All incoming packaging is from virgin sources and consists of three or more types of materials.

If all the following conditions apply, the matrix element is a 4:

- For the case where materials are acquired from suppliers: No solid residues result from resource extraction or during production of materials by recycling (example: petroleum).
- For the case where supplier components/subsystems are used: None/minimal packaging material is used or supplier takes back all packaging material.
- For the case where supplier components/subsystems are used: Incoming packaging is totally reused/recycled.

If neither of the preceding ratings is assigned, complete the checklist below. Assign a rating of 1, 2, or 3 depending on the degree to which the product meets DFE preferences for this matrix element.

- Is the product designed to minimize the use of materials whose extraction or purification involves the production of large amounts of solid residues (i.e., coal and all virgin metals)?
- Is the product designed to minimize the use of materials whose extraction or purification involves the production of toxic solid residues? (This category includes all radioactive materials.)
- Has incoming packaging volume and weight, at and among all levels (primary, secondary, and tertiary), been minimized?
- Is materials diversity minimized in incoming packaging?

PRODUCT MATRIX ELEMENT 1, 4

Life Stage: Premanufacture

Environmental Stressor: Liquid Residues

If either of the following conditions apply, the matrix element rating is 0:

- For the case where supplier components/subsystems are used: Metals from virgin ores that cause substantial acid mine drainage are used, and suitable virgin material is available from recycling streams. (Materials causing acid mine drainage are defined as copper, iron, nickel, lead, and zinc.)
- For the case where materials are acquired from suppliers: The packaging contains toxic or hazardous substances that might leak from it if improper disposal occurs.

If the both of the following conditions apply, the matrix element is a 4:

- For the case where materials are acquired from suppliers: No liquid residues result from resource extraction or during production of materials by recycling.
- For the case where supplier components/subsystems are used: No liquid residue is generated during transportation, unpacking, or use of this product.

If neither of the preceding ratings is assigned, complete the checklist below. Assign a rating of 1, 2, or 3 depending on the degree to which the product meets DFE preferences for this matrix element.

- Is the product designed to minimize the use of materials whose extraction or purification involves the generation of large amounts of liquid residues? (This category includes paper and allied products, coal, and materials from biomass.)
- Is the product designed to minimize the use of materials whose extraction or purification involves the generation of toxic liquid residues? (These materials are defined as aluminum, chromium, copper, iron, lead, mercury, nickel, and zinc.)
- Are refillable/reusable containers used for incoming liquid materials where appropriate?
- Does the use of incoming components require cleaning that involve a large amount of water or that generate liquid residues needing special disposal methods?

PRODUCT MATRIX ELEMENT 1, 5

Life Stage: Premanufacture

Environmental Stressor: Gaseous Residues

If the following condition applies, the matrix element rating is 0:

- The materials used cause substantial emissions of toxic, smog-producing, or greenhouse gases into the environment, and suitable alternatives that do not do so are available. (These materials are defined as aluminum, chromium, copper, iron, lead, mercury, nickel, zinc, paper and allied products, and concrete.)

If the following condition applies, the matrix element is a 4:

- No gaseous residues are produced during resource extraction or production of materials by recycling.

If neither of the preceding ratings is assigned, complete the checklist below. Assign a rating of 1, 2, or 3 depending on the degree to which the product meets DFE preferences for this matrix element.

- Is the product designed to minimize the use of materials whose extraction or purification involves the production of large amounts of gaseous (toxic or otherwise) residues? (Such materials are defined as aluminum, copper, iron, lead, nickel, and zinc.)

PRODUCT MATRIX ELEMENT 2, 1

Life Stage: Product manufacture

Environmental Stressor: Materials Choice

If the following condition applies, the matrix element rating is 0:

- Product manufacture requires relatively large amounts of materials that are restricted (see [1,1]), toxic, and/or radioactive.

If the following condition applies, the matrix element rating is a 4:

- Materials used in manufacture are completely closed loop (captured and reused/recycled) with minimum inputs required.

If neither of the preceding ratings is assigned, complete the checklist below. Assign a rating of 1, 2, or 3 depending on the degree to which the product meets DFE preferences for this matrix element.

- Do manufacturing processes avoid the use of materials that are in restricted supply?
- Is the use of toxic material avoided or minimized?
- Is the use of radioactive material avoided?
- Is the use of virgin material minimized?
- Has the chemical treatment of materials and components been minimized?

PRODUCT MATRIX ELEMENT 2, 2

Life Stage: Product manufacture

Environmental Stressor: Energy Use

If the following condition applies, the matrix element rating is 0:

- Energy use for product manufacture/testing is high and less energy-intensive alternatives are available.

If the following condition applies, the matrix element is a 4:

- Product manufacture/testing requires no or minimal energy use.

If neither of the preceding ratings is assigned, complete the checklist below. Assign a rating of 1, 2, or 3 depending on the degree to which the product meets DFE preferences for this matrix element.

- Is the product manufacture designed to minimize the use of energy-intensive processing steps?
- Is the product manufacture designed to minimize energy-intensive evaluation/testing steps?
- Do the manufacturing processes use co-generation, heat exchanges, and/or other techniques to utilize otherwise wasted energy?
- Is the manufacturing facility powered down when not in use?

PRODUCT MATRIX ELEMENT 2, 3

Life Stage: Product manufacture

Environmental Stressor: Solid Residues

If the following condition applies, the matrix element rating is 0:

- Solid manufacturing residues are large and no reuse/recycling programs are in use.

If the following condition applies, the matrix element is a 4:

- Solid manufacturing residues are minor and each constituent is >90 percent reused/recycled.

If neither of the preceding ratings is assigned, complete the checklist below. Assign a rating of 1, 2, or 3 depending on the degree to which the product meets DFE preferences for this matrix element.

- Have solid manufacturing residues been minimized and reused to the greatest extent possible?
- Has the resale of all solid residues as inputs to other products/processes been investigated and implemented?
- Are solid manufacturing residues that do not have resale value minimized and recycled?

PRODUCT MATRIX ELEMENT 2, 4

Life Stage: Product manufacture

Environmental Stressor: Liquid Residues

If the following condition applies, the matrix element rating is 0:

- Liquid manufacturing residues are large and no reuse/recycling programs are in use.

If the following condition applies, the matrix element is a 4:

- Liquid manufacturing residues are minor and each constituent is >90 percent reused/recycled.

If neither of the preceding ratings is assigned, complete the checklist below. Assign a rating of 1, 2, or 3 depending on the degree to which the product meets DFE preferences for this matrix element.

- If solvents or oils are used in the manufacture of this product, is their use minimized and have alternatives been investigated and implemented?
- Have opportunities for sale of all liquid residues as input to other processes/products been investigated and implemented?
- Have the processes been designed to require maximum recycled liquid process chemicals rather than virgin materials?

PRODUCT MATRIX ELEMENT 2, 5

Life Stage: Product manufacture

Environmental Stressor: Gaseous Residues

If either of the following conditions apply, the matrix element rating is 0:

- Gaseous manufacturing residues are large and no reuse/recycling programs are in use.
- CFCs are used in product manufacture.

If the following condition applies, the matrix element is a 4:

- Gaseous manufacturing residues are relatively minor and reuse/recycling programs are in use.

If neither of the preceding ratings is assigned, complete the checklist below. Assign a rating of 1, 2, or 3 depending on the degree to which the product meets DFE preferences for this matrix element.

- If HCFCs are used in the manufacture of this product, have alternatives been thoroughly investigated and implemented?
- Are greenhouse gases used or generated in any manufacturing process connected with this product?
- Have the resale of all gaseous residues as inputs to other processes/products been investigated?

PRODUCT MATRIX ELEMENT 3, 1

Life Stage: Product packaging and transport

Environmental Stressor: Materials Choice

If the following condition applies, the matrix element rating is 0:

- All outgoing packaging is from virgin sources and consists of three or more types of materials.

If the following condition applies, the matrix element rating is a 4:

- No outgoing packaging or minimal and recycled packaging material is used.

If neither of the preceding ratings is assigned, complete the checklist below. Assign a rating of 1, 2, or 3 depending on the degree to which the product meets DFE preferences for this matrix element.

- Does the product packaging minimize the number of different materials used and is it optimized for weight/volume efficiency?
- Have efforts been made to use recycled materials for product packaging and to make sure the resulting package is recyclable and marked as such?
- Have the packaging engineer and the installation personnel been consulted during product design?

PRODUCT MATRIX ELEMENT 3, 2

Life Stage: Product packaging and transportation

Environmental Stressor: Energy Use

If any of the following condition applies, the matrix element rating is 0:

- Packaging material extraction, packaging procedure, and transportation/installation method(s) are all energy intensive and less energy-intensive options are available.

If the following condition applies, the matrix element is a 4:

- Packaging material extraction, packaging procedure, and transportation/installation method(s) all require little or no energy.

If neither of the preceding ratings is assigned, complete the checklist below. Assign a rating of 1, 2, or 3 depending on the degree to which the product meets DFE preferences for this matrix element.

- Do packaging procedures avoid energy-intensive activities?
- Are component supply systems and product distribution/installation plans designed to minimize energy use?
- If installation is involved, is it designed to avoid energy-intensive procedures?
- Is long distance, energy-intensive product transportation avoided or minimized?

PRODUCT MATRIX ELEMENT 3, 3

Life Stage: Product packaging and transportation

Environmental Stressor: Solid Residues

If the following condition applies, the matrix element rating is 0:

- Outgoing packaging material is excessive, with little consideration given to recycling or reuse.

If the following condition applies, the matrix element is a 4:

- Minimal or no outgoing packaging material is used and/or the packaging is totally reused or recycled.

If neither of the preceding ratings is assigned, complete the checklist below. Assign a rating of 1, 2, or 3 depending on the degree to which the product meets DFE preferences for this matrix element.

- Is the product packaging designed to make it easy to separate the constituent materials?
- Do the packaging materials need special disposal after products are unpacked?
- Has product packaging volume and weight, at and among all three levels (primary, secondary, and tertiary) been minimized?

- Are arrangements made to take back product packaging for reuse and/or recycling?
- Is materials diversity minimized in outgoing product packaging?

PRODUCT MATRIX ELEMENT 3, 4

Life Stage: Product packaging and transportation
Environmental Stressor: Liquid Residues

If the following condition applies, the matrix element rating is 0:

- The product packaging contains toxic or hazardous substances that might leak from it if improper disposal occurs (such as acid from batteries).

If the following condition applies, the matrix element is a 4:

- Little or no liquid residue is generated during packaging, transportation, or installation of this product.

If neither of the preceding ratings is assigned, complete the checklist below. Assign a rating of 1, 2, or 3 depending on the degree to which the product meets DFE preferences for this matrix element.

- Are refillable or reusable containers used for liquid products where appropriate?
- Do the product packaging operations need cleaning/maintenance procedures that require a large amount of water or generate other liquid residues (oils, detergents, etc.) that need special methods of disposal?
- Do the product unpacking and/or installation operations require cleaning that involves large amounts of water or that generate liquid residues needing special disposal methods?

PRODUCT MATRIX ELEMENT 3, 5

Life Stage: Product packaging and transportation
Environmental Stressor: Gaseous Residues

If the following condition applies, the matrix element rating is 0:

- Abundant gaseous residues are generated during packaging, transportation, or installation, and alternative methods that would significantly reduce gaseous emissions are available.

If the following condition applies, the matrix element is a 4:

- Little or no gaseous residues are generated during packaging, transportation, or installation of this product.

If neither of the preceding ratings is assigned, complete the checklist below. Assign a rating of 1, 2, or 3 depending on the degree to which the product meets DFE preferences for this matrix element.

- If the product contains pressurized gases, are transport/installation procedures designed to avoid their release?
- Are product distribution plans designed to minimize gaseous emissions from transport vehicles?
- If the packaging is recycled for its energy content (i.e., incinerated), have the materials been selected to ensure that no toxic gases are released?

PRODUCT MATRIX ELEMENT 4, 1

Life Stage: Product use
Environmental Stressor: Materials Choice

If the following condition applies, the matrix element rating is 0:

- Consumables contain significant quantities of materials in restricted supply or toxic/hazardous substances.

If the following condition applies, the matrix element rating is a 4:

- Product use and product maintenance require no consumables.

If neither of the preceding ratings is assigned, complete the checklist below. Assign a rating of 1, 2, or 3 depending on the degree to which the product meets DFE preferences for this matrix element.

- Has consumable material use been minimized?
- If the product is designed to be disposed of after using, have alternative approaches for accomplishing the same purpose been examined?
- Have the materials been chosen such that no environmentally inappropriate maintenance is required, and no unintentional release of toxic materials to the environment occurs during use?
- Are consumable materials generated from recycled streams rather than virgin material?

PRODUCT MATRIX ELEMENT 4, 2

Life Stage: Product use
Environmental Stressor: Energy Use

If any of the following condition applies, the matrix element rating is 0:

- Product use and/or maintenance is relatively energy intensive and less energy-intensive methods are available to accomplish the same purpose.

If the following condition applies, the matrix element is a 4:

- Product use and maintenance requires little or no energy.

If neither of the preceding ratings is assigned, complete the checklist below. Assign a rating of 1, 2, or 3 depending on the degree to which the product meets DFE preferences for this matrix element.

- Has the product been designed to minimize energy use while in service?
- Has energy use during maintenance/repair been minimized?
- Have energy-conserving design features (such as auto shutoff or enhanced insulation) been incorporated?
- Can the product monitor and display its energy use and/or its operating energy efficiency while in service?

PRODUCT MATRIX ELEMENT 4, 3

Life Stage: Product use

Environmental Stressor: Solid Residues

If the following condition applies, the matrix element rating is 0:

- Product generates significant quantities of hazardous/toxic solid residue during use or from repair/maintenance operations.

If the following condition applies, the matrix element is a 4:

- Product generates no (or relative minor amounts of) solid residue during use or from repair/maintenance operations.

If neither of the preceding ratings is assigned, complete the checklist below. Assign a rating of 1, 2, or 3 depending on the degree to which the product meets DFE preferences for this matrix element.

- Has the periodic disposal of solid materials (such as cartridges, containers, or batteries) associated with the use and/or maintenance of this product been avoided or minimized?
- Have alternatives to the use of solid consumables been thoroughly investigated and implemented where appropriate?
- If intentional dissipative emissions to land occur as a result of using this product, have less environmentally harmful alternatives been investigated?

PRODUCT MATRIX ELEMENT 4, 4

Life Stage: Product use

Environmental Stressor: Liquid Residues

If the following condition applies, the matrix element rating is 0:

- Product generates significant quantities of hazardous/toxic liquid residue during use or from repair/maintenance operations.

If the following condition applies, the matrix element is a 4:

- Product generates no (or relatively minor amounts of) liquid residue during use or from repair/maintenance operations.

If neither of the preceding ratings is assigned, complete the checklist below. Assign a rating of 1, 2, or 3 depending on the degree to which the product meets DFE preferences for this matrix element.

- Has the periodic disposal of liquid materials (such as lubricants and hydraulic fluid) associated with the use and/or maintenance of this product been avoided or minimized?
- Have alternatives to the use of liquid consumables been thoroughly investigated and implemented where appropriate?
- If intentional dissipative emissions to land occur as a result of using this product, have less environmentally harmful alternatives been investigated?
- If product contains liquid material that has the potential to be unintentionally dissipated during use or repair, have appropriate preventive measures been incorporated?

PRODUCT MATRIX ELEMENT 4, 5

Life Stage: Product use

Environmental Stressor: Gaseous Residues

If the following condition applies, the matrix element rating is 0:

- Product generates significant quantities of hazardous/toxic gaseous residue during use or from repair/maintenance operations.

If the following condition applies, the matrix element is a 4:

- Product generates no (or relative minor amounts of) gaseous residue during use or from repair/maintenance operations.

If neither of the preceding ratings is assigned, complete the checklist below. Assign a rating of 1, 2, or 3 depending on the degree to which the product meets DFE preferences for this matrix element.

- Has the periodic disposal of liquid materials (such as CO_2, SO_2, VOCs, and CFCs) associated with the use and/or maintenance of this product been avoided or minimized?
- Have alternatives to the use of gaseous consumables been thoroughly investigated and implemented where appropriate?
- If intentional dissipative emissions to land occur as a result of using this product, have less environmentally harmful alternatives been investigated?
- If product contains gaseous materials that have the potential to be unintentionally dissipated during use or repair, have appropriate preventive measures been incorporated?

PRODUCT MATRIX ELEMENT 5, 1

Life Stage: Disposal

Environmental Stressor: Materials Choice

If the following condition applies, the matrix element rating is 0:

- Product contains significant quantities of mercury (i.e., mercury relays), asbestos (i.e., asbestos based insulations), or cadmium (i.e., cadmium or zinc plated parts) that are not clearly identified and easily removed.

If the following condition applies, the matrix element rating is a 4:

- Material diversity is minimized, the product is easy to disassemble, and all parts are recyclable.

If neither of the preceding ratings is assigned, complete the checklist below. Assign a rating of 1, 2, or 3 depending on the degree to which the product meets DFE preferences for this matrix element.

- Have materials been chosen and used in light of the desired recycling/disposal option for the product (e.g., for incineration, for recycling, for refurbishment)?
- Does the product minimize the number of different materials that are used in its manufacture?
- Are the different materials easy to identify and separate?
- Is this a battery free product?
- Is this product free of components containing PCBs or PCTs (e.g., in-capacitors and transformers)?
- Are major plastics parts free of polybrominated flame retardants or heavy metal-based additives (colorants, conductors, stabilizers, etc.)?

PRODUCT MATRIX ELEMENT 5, 2

Life Stage: Disposal

Environmental Stressor: Energy Use

If any of the following condition applies, the matrix element rating is 0:

- Recycling/disposal of this product is relative energy intensive (compared to other products that perform the same function) due to its weight, construction, and/or complexity.

If the following condition applies, the matrix element is a 4:

- Energy use for recycling or disposal of this product is minimal.

If neither of the preceding ratings is assigned, complete the checklist below. Assign a rating of 1, 2, or 3 depending on the degree to which the product meets DFE preferences for this matrix element.

- Is the product designed with the aim of minimizing the use of energy-intensive process steps in disassembly?
- Is the product designed for high-level reuse of materials? (Direct reuse in a similar product is preferable to degraded reuse).
- Will transport of products for recycling be energy intensive because of product weight or volume or the location of the recycling facilities?

PRODUCT MATRIX ELEMENT 5, 3

Life Stage: Disposal

Environmental Stressor: Solid Residues

If the following condition applies, the matrix element rating is 0:

- Product consists primarily of unrecyclable solid materials (such as rubber, fiberglass, and compound polymers).

If the following condition applies, the matrix element is a 4:

- Product can be easily refurbished and reused and is easily dismantled and 100 percent reused/recycled at the end of its life. For example, no part of this product will end up in a landfill.

If neither of the preceding ratings is assigned, complete the checklist below. Assign a rating of 1, 2, or 3 depending on the degree to which the product meets DFE preferences for this matrix element.

- Has the product been assembled with fasteners such as clips or hook-and-loop attachments rather than chemical bonds (gels, potting compounds) or welds?
- Have efforts been made to avoid joining dissimilar materials together in ways difficult to reverse?
- Are all plastic components identified by ISO markings as to their content?
- If product consists of plastic parts, is there one dominant (>80 percent by weight) species?
- Is this product to be leased rather than sold?

PRODUCT MATRIX ELEMENT 5, 4

Life Stage: Disposal

Environmental Stressor: Liquid Residues

If the following condition applies, the matrix element rating is 0:

- Product contains primarily unrecyclable liquid materials.

If the following condition applies, the matrix element is a 4:

- Product uses no operating liquids (such as oils, coolants, or hydraulic fluids) and no cleaning agents or solvents are necessary for its reconditioning.

If neither of the preceding ratings is assigned, complete the checklist below. Assign a rating of 1, 2, or 3 depending on the degree to which the product meets DFE preferences for this matrix element.

- Can liquids contained in the product be recovered at disassembly rather than lost?
- Does disassembly, recovery, and reuse generate liquid residues?
- Does materials recovery and reuse generate liquid residues?

PRODUCT MATRIX ELEMENT 5, 5

Life Stage: Disposal

Environmental Stressor: Gaseous Residues

If the following condition applies, the matrix element rating is 0:

- Product contains or produces primarily unrecyclable gaseous materials that are dissipated to the atmosphere at the end of its life.

If the following condition applies, the matrix element is a 4:

- Product contains no substances lost to evaporation/sublimation (other than water) and no volatile substances are used for refurbishment.

If neither of the preceding ratings is assigned, complete the checklist below. Assign a rating of 1, 2, or 3 depending on the degree to which the product meets DFE preferences for this matrix element.

- Can gases contained in the product be recovered at disassembly rather than lost?
- Does materials recovery and reuse generate gaseous residues?
- Can plastic parts be incinerated without requiring sophisticated air pollution control devices? Plastic parts that can cause difficulty in this regard are those that contain polybrominated flame retardants or metal-based additives, are finished with polyurethane-based paints, or are plated or painted with metals.

Appendix D

Concept Comparison Using Cost and Environmental Considerations

In this section, we will discuss the usage of the concept comparison matrix shown in Table 8.5 in more detail. There are several preliminary steps that must occur before this table can be assembled. First, either a house of quality must be developed for the product (discussed in Appendix A), or at least a list of critical product criteria must be gathered and these criteria given an importance score on a scale of 1-10, where 10 is most important. Next, a streamlined life cycle assessment should be performed. We modified the version shown in Appendix C, and developed the SLCA of the safe shown in Table 8.4. In this table, each life cycle stage is rated on a scale of 0-4 with 0 being of extreme concern and 4 being of no concern. We followed the rating scheme discussed in Appendix C. Finally, cost criteria were developed and included: raw material costs, capital equipment costs, waste disposal costs, and labor costs.

We can now begin to develop the concept comparison matrix. A weight is assigned by the design team for the areas of quality, environment, and cost. In this example, quality and cost are weighted as 0.4 and environment as 0.2. Next the current processes and the two concepts are listed. We are interested in comparing powder coating to replace the current spray process and recycling process in the cleaning station to recycle the wash water. Next the criteria for judging these concepts are listed across the top. We use the following quality criteria: aesthetics of the paint finish, safe security, and fire proofing. All three of these criteria are important in safe performance, and are given raw importance scores of 9, 5, and 6, respectively. From the SLCA matrix shown in Table 8.4, the most critical elements are solid, liquid, and gaseous residues or waste, health hazards, and environmental externalities, which in this case is the waste hazard. Because the scales are different between the SLCA and the concept comparison matrix, we have to translate the importance scores. For example, the liquid waste for the paint process and the wash cycle was rated a 1 (significant concern) in the SLCA, which was translated to a 7 in the comparison matrix since the scale runs from 4 to 0 in the

SLCA and from 1-10 in the comparison matrix. For the case of health concerns or hazards, the paint line was of extreme concern and the wash station was of minor concern, so in the comparison matrix it was given an importance weight of 9. Finally, the cost criteria importance weights are assigned by the design team with waste disposal costs and capital equipment costs both receiving a 7, and with raw material and labor costs receiving 3s. Next, these importance score are normalized for each category. For example, in the quality category, the raw scores are 9, 5, and 6, which are summed and then each score is divided by the sum and multiplied by 100 to get 45, 25, and 30, respectively. Finally, the overall weight is calculated by multiplying the normalized weight by the team-assigned weight. For example, in the aesthetics column, the normalized weight of 45 is multiplied by 0.4 to get 18. This is done for each criterion.

Now we can complete the rows at the top. The current processes that are used are listed along with the alternatives being considered. In this case, the current spray painting process is compared with powder coating and the wash station is compared with one using recycling system. Each design alternative is then rated on a scale of 1-10 on how well it meets the criteria with 1 not meeting and 10 meeting completely. For example, for the aesthetics (paint finish) criterion, the current spray process received a 9 and the powder coating process received a 7 since its finish is not mirror quality where the spray has a much glossier finish. For the wash station, the aesthetics remain the same and rate a 5. The satisfaction is calculated by multiplying the concept score by the overall weight and summing across the row. For this example we have the following satisfaction score for the VOC-based spray paint:

$$9(18)+7(10)+5(12)+5(4)+3(2.9)+1(4)+3(4)+1(5.1)+5(14)+5(6)+5(14)+1(6)=518$$

Finally, the percent difference is calculated. From this table we can see that there is a 30% improvement in score for the powder coating process versus a 3% for the recycling system. Therefore, we can easily see that powder coating is worth the capital investment whereas the recycling system does not have the same payoff.

Index

D

P

Q

R

S